Chapman & Hall/CRC Mathematical and Computational Biology Series

Niche Modeling

Predictions from Statistical Distributions

CHAPMAN & HALL/CRC
Mathematical and Computational Biology Series

Aims and scope:

This series aims to capture new developments and summarize what is known over the whole spectrum of mathematical and computational biology and medicine. It seeks to encourage the integration of mathematical, statistical and computational methods into biology by publishing a broad range of textbooks, reference works and handbooks. The titles included in the series are meant to appeal to students, researchers and professionals in the mathematical, statistical and computational sciences, fundamental biology and bioengineering, as well as interdisciplinary researchers involved in the field. The inclusion of concrete examples and applications, and programming techniques and examples, is highly encouraged.

Series Editors

Alison M. Etheridge
Department of Statistics
University of Oxford

Louis J. Gross
Department of Ecology and Evolutionary Biology
University of Tennessee

Suzanne Lenhart
Department of Mathematics
University of Tennessee

Philip K. Maini
Mathematical Institute
University of Oxford

Shoba Ranganathan
Research Institute of Biotechnology
Macquarie University

Hershel M. Safer
Weizmann Institute of Science
Bioinformatics & Bio Computing

Eberhard O. Voit
The Wallace H. Couter Department of Biomedical Engineering
Georgia Tech and Emory University

Proposals for the series should be submitted to one of the series editors above or directly to:
CRC Press, Taylor & Francis Group
24-25 Blades Court
Deodar Road
London SW15 2NU
UK

Published Titles

Cancer Modelling and Simulation
Luigi Preziosi

Computational Biology: A Statistical Mechanics Perspective
Ralf Blossey

Computational Neuroscience: A Comprehensive Approach
Jianfeng Feng

Data Analysis Tools for DNA Microarrays
Sorin Draghici

Differential Equations and Mathematical Biology
D.S. Jones and B.D. Sleeman

Exactly Solvable Models of Biological Invasion
Sergei V. Petrovskii and Bai-Lian Li

Introduction to Bioinformatics
Anna Tramontano

An Introduction to Systems Biology: Design Principles of Biological Circuits
Uri Alon

Knowledge Discovery in Proteomics
Igor Jurisica and Dennis Wigle

Modeling and Simulation of Capsules and Biological Cells
C. Pozrikidis

Niche Modeling: Predictions from Statistical Distributions
David Stockwell

Normal Mode Analysis: Theory and Applications to Biological and Chemical Systems
Qiang Cui and Ivet Bahar

Stochastic Modelling for Systems Biology
Darren J. Wilkinson

The Ten Most Wanted Solutions in Protein Bioinformatics
Anna Tramontano

Chapman & Hall/CRC Mathematical and Computational Biology Series

Niche Modeling
Predictions from Statistical Distributions

David Stockwell

CRC Press
Taylor & Francis Group
Boca Raton London New York

CRC Press is an imprint of the
Taylor & Francis Group, an **informa** business
A CHAPMAN & HALL BOOK

CRC Press
Taylor & Francis Group
6000 Broken Sound Parkway NW, Suite 300
Boca Raton, FL 33487-2742

First issued in paperback 2019

ISBN-13: 978-1-58488-494-1 (hbk)
ISBN-13: 978-0-367-38970-3 (pbk)

Library of Congress Cataloging-in-Publication Data

Stockwell, David R. B. (David Russell Bancroft)
   Ecological niche modeling : ecoinformatics in application to biodiversity / David R.B. Stockwell.
      p. cm. -- (Mathematical and computational biology series)
   Includes bibliographical references.
   ISBN-13: 978-1-58488-494-1 (alk. paper)
   ISBN-10: 1-58488-494-0 (alk. paper)
   1. Niche (Ecology)--Mathematical models. 2. Niche (Ecology)--Computer simulation. I. Title. II. Series.

QH546.3.S76 2006
577.8'2--dc22   2006027353

Visit the Taylor & Francis Web site at
http://www.taylorandfrancis.com

and the CRC Press Web site at
http://www.crcpress.com

Contents

List of Tables

List of Figures

0.1 Preface

Niche modeling is a relatively new field of research aimed at helping us to understand the response of species to their environment and predicting their distribution. The practice of niche modeling uses tools from mathematics and statistics, data management and geographic spatial analysis. The first six chapters are concerned with fundamentals, programming, theory and examples of niche modeling. When used in conjunction with more detailed and specific texts and manuals, students and researchers may successfully do niche modeling for the first time.

Successful niche modeling also requires an understanding of the limitations and potential pitfalls of prediction. Due to the importance of avoiding errors, the last six chapters are devoted to sources of errors. All are relatively novel topics in the field: autocorrelation, bias, long term persistence, non-linearity, circularity and fraud, and should be of interest to researchers.

While a statistical language like R or S-plus is not essential, it provides a way of describing these main concepts, showing someone how to use them, and hands on experience at solving problems through examples. It is assumed that readers have a basic knowledge of mathematics and programming.

Above all, successful niche modeling requires deep understanding of the process of creating and using probability distributions in multidimensional spatial and temporal application. Here simplified examples complement the rigor and completeness that can be found in the literature. The generality of the approach is illustrated by examples as diverse as invasive species dynamics, predicting house price increases, and detecting management of data or fraud.

I think there are many advantages in developing depth of intuition, such as capacity to develop novel approaches, and avoiding gross errors. Off-the-shelf statistical packages are tailored exactly to applications but can hide problematic complexity. Recipe book implementations fail to educate users in the details, assumptions and pitfalls of the analysis. As each situation is a little different, packages may not be able to adapt to the specific need of their study. Understanding of the basics, and the pitfalls, also creates confidence for communicating the results.

0.1.1 Summary of chapters

1. **Functions** This chapter summarizes major mathematical types, operations and relationships encountered both in the book and in niche modeling. This and the following two chapters could be treated as a tutorial in the R language. For example, the main functions for representing the

inverted U shape characteristic of a niche – step, Gaussian, quadratic and ramp functions – are illustrated both graphically and in R code. The chapter concludes with the ACF and lag plots, in one or two dimensions.

2. **Data** This chapter shows a simple biodiversity database using R. By using data frames as tables, it is possible to replicate the basic spreadsheet and relational database operations with R's powerful indexing functions, eliminating conversion problems as data is moved between systems while learning more about R.

3. **Spatial** R and image processing operations can perform many of the elementary spatial operations necessary for niche modeling. While these do not replace a GIS, it demonstrates generalization of arithmetic concepts to images and efficient implementation of simple spatial operations.

4. **Topology** Set theory helps to identify the basic assumptions underlying niche modeling, and the relationships and constraints between these assumptions. The chapter shows the standard definition of the niche as environmental envelopes around all ecologically relevant variables is equivalent to a box topology. A proof is offered that the Hutchinsonian environmental envelope definition of a niche when extended to large or infinite dimensions of environmental variables loses desirable topological properties. This argues for the necessity of careful selection of a small set of environmental variables.

5. **Environmental data collections** Management of data for niche modeling is poorly served by user-developed files stored in a local directory. A wide variety of data sets are currently available, and better quality niche modeling will result from using data in true archives – shared by many studies and trusted with the highest level of quality. A number of sources of data are described and access issues discussed.

6. **Examples** The three examples of niche models here were selected to contradict three main misconceptions of niche modeling. The house price increase example shows a niche that is bimodal and not an inverted U. The second example of the Brown Treesnake shows an asymptotic response with respect to precipitation. The third example of the zebra mussel shows how dynamic models of the spread of invasive species can be developed from the niche model, contrary to the view that niche models are restricted to equilibrium approaches.

7. **Bias** Here a simple theoretical model of range-shift is used to estimate the magnitude of potential bias in estimates of changes in range area due to climate change.

8. **Autocorrelation** This chapter shows the problem of validating models on autocorrelated data using internal or external validation. Holding

back data at random is shown to be inadequate to determine the skill of a model when the data are autocorrelated, particularly when using smoothed data.

9. **Nonlinearity** Procedures with linear assumptions are not reliable when the responses are non-linear. Here using simulations and a linear model for reconstructing past temperatures, niche model-like tree responses create artifacts including signal degradation, loss of variance, temporal shifts in peaks, and period doubling.

10. **Long Term Persistence** The natural world is more uncertain and more indeterministic than modeled using classical statistics. Here we show evidence that temporal and spatial natural series display LTP, or scale invariant distributions. These results provide no justification for models with preferred spatial or temporal scale, which greatly underestimate confidence limits.

11. **Circularity** A major source of error is due to conclusions encoded into the assumptions of the methodology, so allowing no other conclusion than the one obtained. Here we show a potential approach to the problem of quantifying circular reasoning. By feeding random data with the same noise and autocorrelation properties into a methodology, one obtains a null model with benchmarks for rejection regions, and expectations incorporating hidden model assumptions.

12. **Fraud** The accidental or fraudulent management of results can be detected using the distributional modeling methods of niche modeling. The second digit distribution postulated by Benford's Law allows detection of fabricated data in natural time series drawn from a single distribution. The approach is applied to a range of natural data.

I would like to express my thanks to providers of data used to illustrate issues in niche modeling. The Brown Treesnake point data were from a listing of the Australian Museum holdings provided by Gordon Rodda. Zebra Mussel occurrence data were provided by Amy J. Benson. Temperature reconstruction data were provided by Steve McIntyre. Thank you also to the San Diego Supercomputer Center, University of California San Diego, and to the National Center for Ecological Analysis and Synthesis, University of California Santa Barbara, for providing financial support and office space, funded under a sabbatical research program by the United States National Science Foundation. The development and refinement of some of the sections of the book were assisted by exchanges via a weblog. Steve McIntyre, Demetris Koutsoyiannis, Martin Ringo, and anonymous correspondent TCO were particularly helpful. I would also like to express my deep appreciation for my wife Siriluck and two children, Lena and Victoria.

Chapter 1

Functions

This chapter summarizes some of the major mathematical and statistical concepts used in niche modeling. The examples illustrate the use of R language, a powerful, reliable and free statistical program [R D05].

1.1 Elements

R is a very powerful language for a number of reasons: particularly vector processing, indexing and function definitions. These allow code to be shortened considerably, loops implemented efficiently, and encourages a parsimonious style of programming around larger data structures that suits statistical scripts.

In approaching R one finds the basic constructs from most programming languages. R supports the basic data types: integer, numeric, logical, character/string. To these R adds advanced types: factor, complex, and raw, and complex containers such as lists, vectors and matrices as follows:

1.1.1 Factor

Factors express ordered or unordered categories and consist of a finite set of named ordered or unordered levels. Factors are the default type R imports into data tables. This can be confusing when you expect numbers. The example shows factors of population density of a species.

```
> factor(c("1", "2", "3", "4"), ordered = TRUE)
```

```
[1] 1 2 3 4
Levels: 1 < 2 < 3 < 4
```

1.1.2 Complex

Complex numbers have the form $x + yi$ where x (the real part) and y (the imaginary part) are real numbers and i the square root of -1. These are a useful type as the two parts can be manipulated as a single number, instead of having to create a more complex type. For example, the two parts can represent the coordinates of a point in a plane.

```
> j <- 154.1 - (0+22.3i)
> x <- 1:30
> x
```

```
[1]  1  2  3  4  5  6  7  8  9 10 11 12 13 14 15 16 17 18 19
[20] 20 21 22 23 24 25 26 27 28 29 30
```

1.1.3 Raw

Type Raw holds raw bytes. The only valid operations on the type raw are the bitwise operations, AND, OR and NOT. Raw values are displayed in hex notation, where the basic digits from 0 to 15 are represented by letters 0 to f.

Raw values are most frequently used in images where the numbers represent intensity, e.g. 255 for white and 0 for black. Raw values can store the categories of vegetation types in a vegetation map or the normalized values of such variables as average temperature or rainfall.

1.1.4 Vectors

Vectors are an ordered set of items of identical type and are one of the most versatile features of R. Below are some of the most common ways of creating a vector:

```
> x <- c(2, 1.5, 4.99, 60.58, 0.05, 3, 12.95, 0.02)
> x
```

```
[1]  2.00  1.50  4.99 60.58  0.05  3.00 12.95  0.02
```

```
> y <- 1:8
> y
```

```
[1] 1 2 3 4 5 6 7 8
```

```
> z <- seq(1982, 1989, by = 1)
> z
```

```
[1] 1982 1983 1984 1985 1986 1987 1988 1989
```

1.1.5 Lists

Lists contain an unordered set of named items of different type. These are a general purpose type for holding all kinds of data. An example of a list below uses a vector of locations of a species and the species name.

```
> list(coords = c(123.12 - (0+45i), 122 - (0+41i),
+       130 - (0+40i)), species = "Puma concolor")

$coords
[1] 123.12-45i 122.00-41i 130.00-40i

$species
[1] "Puma concolor"
```

1.1.6 Data frames

Data frames are an extremely useful construct for organizing data in R, very similar to tables or spreadsheets. A data frame is essentially a list of vectors of equal length. That is, while each column in a table can be a different type, they must all have the same number of items. The *data.frame* command creates data frames, but another common method of creation is by reading in data from files via the *read.table* command. R has a built-in spreadsheet application for editing data.frames called with the *edit* command (Table 1.1).

```
> d <- data.frame(Cost = x, Code = y, Year = z)
```

1.1.7 Time series

Time series are another useful complex construct. Time series allow the elements of a vector to be described along with start dates, and sampling frequencies. They will then be lined-up correctly by the elementary operations. The *ts* command creates a time series.

```
> ts(1:10)

Time Series:
Start = 1
End = 10
```

Niche Modeling

TABLE 1.1:   R contains a spreadsheet-like data editor called with the *edit* command.

	Cost	Code	Year
1	2.00	1.00	1982.00
2	1.50	2.00	1983.00
3	4.99	3.00	1984.00
4	60.58	4.00	1985.00
5	0.05	5.00	1986.00
6	3.00	6.00	1987.00
7	12.95	7.00	1988.00
8	0.02	8.00	1989.00

```
Frequency = 1
 [1]  1  2  3  4  5  6  7  8  9 10
```

1.1.8   Matrix

Below is an example of a matrix of random numbers. Elementary matrix algebra is possible in R, using the standard operators on numbers and vectors as described previously. Coding algorithms as matrix operations can drastically reduce the number of lines of code, improve clarity and increase efficiency of algorithms.

```
> matrix(rnorm(16), 4, 4)

           [,1]        [,2]        [,3]        [,4]
[1,]   1.2085126   1.4399613 -0.6782351 -0.2068214
[2,]  -0.4676946  -0.6252734  0.8457706 -0.5456283
[3,]  -0.1882097   1.0402726 -0.2805549  0.8075877
[4,]   0.4239560   0.9996605 -0.5231428 -0.2089011
```

Table1.2 lists the basic types in R, and examples follow.

1.2   Operations

The types use the usual operators available in most computer languages (e.g. Table 1.3). R usually casts types into the correct form for the operation,

TABLE 1.2: Some basic types in the R language.

	Examples
Integer	7
Numeric	5.6
Logical	TRUE, FALSE
Character	here
Factor	1
Complex	0+0i
Raw	ff
Constants	pi, NULL, Inf, −Inf, nan, NA
Vectors	1:10, rep(1,10), seq(0,10,1)
Matrices	matrix(0,3,3), array(0, c(3,3))
Lists	list(x=1, y=1)
Data frames	data.frame(x=numeric(10), y=character(10))

e.g. integer + float = float.

TABLE 1.3: Some basic operations in the R language.

	Operators
Numeric	x+y, x−y, x*y, x/y, x
Logical	!x, xy, xy, x|y, x||y, xor(x, y), isTRUE(x)
Bitwise	!, |,
Relational	x<y, x>y, x<=y, x>=y, x==y, x!=y
Assignment	x<−value, value−>x
Accessors	x.y, x[y], pkg::name, pkg:::name
Constructors	x:y, x=y, y model

Being a vector language, R overloads these basic operators to apply to complex built-in types. For example, the vectors constructed using *c*, *seq* and *rep* operations above are combined in the listing below. In a vector language more complex structures such as vectors and lists can be treated as basic types because many of the basic operators apply to them. Here are some rules governing vector operations.

- Basic arithmetic operations like addition are element wise on vectors.

- When vectors are of unequal length they wrap around.

- Logical operations on numeric vectors produce logical vectors.

```
> x + y
```

```
[1]   3.00   3.50   7.99 64.58   5.05   9.00 19.95   8.02

> !y

[1] FALSE FALSE FALSE FALSE FALSE FALSE FALSE FALSE

> x > y

[1]   TRUE FALSE   TRUE   TRUE FALSE FALSE   TRUE FALSE
```

In another example of operations on vectors, bitwise operations on raw values perform spatial operations such as masking out areas of an image in order to exclude them from analysis. For example, the bitwise OR operation below converts all values combined with 255 to 255.

```
> as.raw(255)

[1] ff

> as.raw(15) | as.raw(255)

[1] ff
```

The R code below shows how a mask composed of values of 0 or 255 when combined with another image would leave all combinations with the darkest value of 0 unaltered, while converting all areas with value of 255 to the brightest value of 255 (Figure 1.1).

1.3   Functions

There are many ways to introduce functions. An example of the identity function in R is:

```
> f <- function(x) return(x)
> f("this")

[1] "this"
```

```
> par(mfcol = c(1, 3))
> palette(gray(seq(0, 0.9, len = 30)))
> x <- readBin("../ZM/layer10", what = "raw", n = 122 *
+     52)
> y <- readBin("../ZM/mask10", what = "raw", n = 122 *
+     52)
> z <- x | y
> image(matrix(as.numeric(x), 122, 52), ylim = c(1,
+     0), col = 1:30, sub = "A", labels = F)
> image(matrix(as.numeric(y), 122, 52), ylim = c(1,
+     0), col = 1:30, sub = "B", labels = F)
> image(matrix(as.numeric(z), 122, 52), ylim = c(1,
+     0), col = 1:30, sub = "C", labels = F)
```

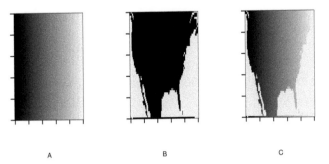

A   B   C

FIGURE 1.1: The bitwise OR combination of two images, A representing longitude and B a mask to give C representing longitude in a masked area.

In mathematical terms, functions are described as a mapping from one domain to another, typically numbers to numbers. More precisely, the mapping f from X to Y is a function provided there is at most one element y of Y related to x via f. The requirement for y to be uniquely determined by the value of x shows that x acts like an index. Based on this definition, indexing of vectors can be regarded as a basic function, where x is the position of the element.

```
> f <- function(x, y) y[x]
> f(3, y)

[1] ff
```

The examples so far have returned single values, but R functions can return more complex return values: vectors, lists and data frames. Many functions operate on whole vectors. The first example below produces a sine wave that could be used to simulate annual temperatures. While the function appears to define the input x as a single value, the result of inputting a twelve element vector is a twelve element sine wave.

```
> daylight <- function(x) -cos(pi * x/6)
> daylight(1:12)

[1] -8.660254e-01 -5.000000e-01 -6.123234e-17  5.000000e-01
[5]  8.660254e-01  1.000000e+00  8.660254e-01  5.000000e-01
[9]  1.836970e-16 -5.000000e-01 -8.660254e-01 -1.000000e+00
```

In the following example the vector indexing function operates on a whole vector.

```
> z <- seq(1981, 1990, by = 1)
> z[1:5]

[1] 1981 1982 1983 1984 1985

> z[z > 1985]

[1] 1986 1987 1988 1989 1990
```

In some cases the syntax of a function differs between vector and unary inputs. Adopting the parallel version allows very compact parallel functions. The example below shows the difference between the simple *max* operation and the parallel *pmax* operation.

```
> max(z, 1984)
```

[1] 1990

```
> pmax(z, 1984)
```

[1] 1984 1984 1984 1984 1985 1986 1987 1988 1989 1990

R also contains a rich set of built in statistical and plotting functions that make many programming tasks simple and efficient. They should be used where possible as they are usually much more efficient than implementations in native R. Table 1.4 lists the main built in functions used throughout this book.

TABLE 1.4:   Some useful built-in functions.

	Description
lm	Fit a linear model
aggregate	Splits the data into subsets, computes summary statistics for each
cumsum	Returns a vector whose elements are the cumulative sums
filter	Applies linear filtering to a univariate time series
hist	Computes and plots a histogram of the given data values
spectrum	Estimates the spectral density of a time series
acf	Computes and plots estimates of the autocorrelation function

1.4   Ecological models

There is a basic distinction between niche modeling and other kinds of biological and ecological models. A species responds to its environment in terms of some species characteristic, such survival, increase in numbers of individuals, or their size or weight. Figure 1.2 shows some basic functions used for describing responses: linear, exponential, power and ramp functions. These functions are the basic way ecologists have described functional response of organisms to such environmental factors as nutrient to predator populations. Typically responses are combinations of linear, and exponential or power relationships.

```
> i <- 1:100
> response <- 1 - i/100
> plot(i, response, type = "l", labels = F)
> lines(i, exp(-i/10), lty = 2)
> lines(i, 10/i, lty = 3)
```

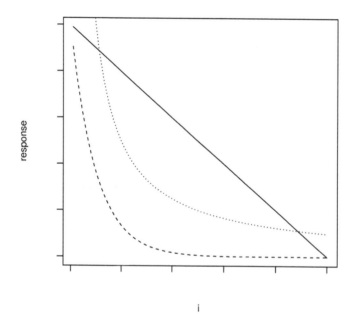

FIGURE 1.2: Basic functions used in modeling: linear, exponential or power relationships.

1.4.1 Preferences

While useful, these functions of response do not adequately express the basic concept of a niche, intimately related to species' preferences.

A niche model, at a minimum, represents the tendency of a species to survive in or prefer a particular set of values. To express this a function requires a 'hump' or inverted U shape centered on values optimal to the species' growth and reproduction.

Figure 1.3 illustrates four such functions:

- the step function,
- a truncated quadratic,
- and exponential, and
- the ramp.

Each of the forms above has slightly different properties. The simple step function expresses the environmental limits of a species. The uniform distribution within the step is less biologically realistic than the truncated quadratic, with preferences declining from the optimum to the extremes. The exponential has long tails like a Gaussian distribution, and so is convenient mathematically, but it suggests some probability of existence, even at great distances from the optimum. The ramp function is often used for computational convenience. Choice of a form may depend on factors such as mathematical tractability as much as realism.

In zero dimensions, responses apply to sets of observations, individuals, or individual survey locations. One-dimensional responses can most often apply over time, e.g. the sales of a product, but also the size of a population over an altitudinal transect.

Periodic responses describe daily, annual, and multiyear cycles. Often periodic relationship are added together, e.g. a weekly and monthly trend leading to more complex functions (Figure 1.4).

1.4.2 Stochastic functions

Stochastic functions incorporate randomness into models. A stochastic function can be defined zero, one or higher dimensionally. For example, given:

$$E_i = f(i)$$

then for when i is an element of a:

Niche Modeling

```
> i <- seq(-2, 2, 0.01)
> preference <- pmax(1 - i^2, 0)
> plot(i, preference, type = "l", labels = F)
> sf <- stepfun(c(-1, 1), c(0, 1, 0), f = 0)
> lines(i, sf(i), lty = 2)
> lines(i, exp(-(i^2)), lty = 3)
> lines(i, c(rep(0, 100), seq(0, 1, 0.01), seq(1,
+      0, -0.01), rep(0, 99)), lty = 4)
```

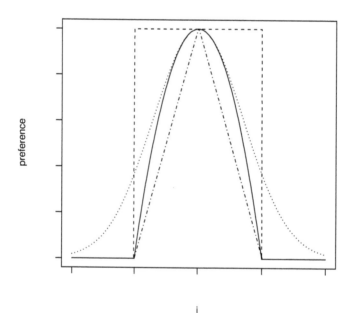

FIGURE 1.3: Basic functions used to represent niche model preference relationships: a step function, a truncated quadratic, exponential and a ramp.

```
> y1 <- sin(pi * (i - 1))
> y2 <- sin(2 * pi * (i - 1))
> periodic <- y1 + y2
> plot(i, periodic, type = "l", labels = F)
> lines(i, y1, lty = 2)
> lines(i, y2, lty = 4)
```

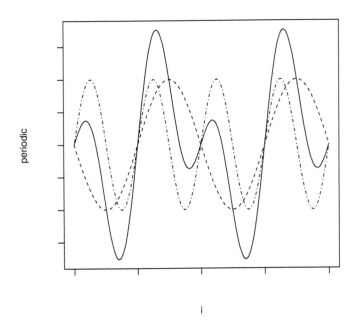

FIGURE 1.4: Cyclical functions are common responses to environmental cycles, both singly and added together to produce more complex patterns.

- set, such as temperature or prices,

- line, e.g. if time then records changes in temperature or prices over time (i.e. a time series), or

- higher dimension, such as a matrix representing a spatial pattern on a landscape.

A stochastic series has an underlying distribution, such as the Gaussian or 'normal' distribution. Stochastic series can also have interrelationships between each of the values, such as autocorrelation. While correlation quantifies the degree to which two variables are related to each other, autocorrelation is the degree to which a variable is related to itself. In positive autocorrelation adjacent values do not differ as much as values that are further apart. In negative autocorrelation, adjacent values differ more than values that are farther apart.

The autocorrelation function (ACF) maps the autocorrelation from the successive distance between points called *lags* to the correlation, from a perfect correlation of plus one to zero or even negative correlations. The ACF is a very useful tool for characterizing and diagnosing the nature of variables.

Examples of stochastic series and their autocorrelation properties are given below.

1.4.2.1   Independent and Identically Distributed

Independent and Identically Distributed (IID) refers to series with independent and identically distributed random numbers. Every y value is a simple random number with no reference to any other number. In R the $rnorm(n)$ function returns a vector of n normally distributed IID random numbers with a mean of zero and a standard deviation of one (Figure 1.5).

The IID series has high autocorrelation at zero lag, as every number correlates perfectly with itself. Correlation at all other lags are below the level of significance, indicated by the dashed line, demonstrating that in an IID series the values are *independent* of each other.

1.4.2.2   Moving Average

Moving Average (MA) series results from applying an averaging function in a moving window over the points in the series (Figure 1.6). As can be seen in the plot of the ACF, the averaging process produces some autocorrelation between neighboring points with low lag. A distinctive feature of series with a moving average is oscillation of the correlation with increasing lags. Some values at each end of the averaged series are consumed by the filter operation.

```
> par(mfrow = c(2, 1))
> IID <- as.ts(rnorm(200))
> plot(IID)
> acf(IID)
```

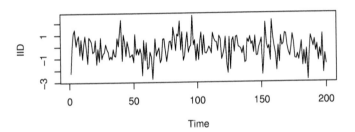

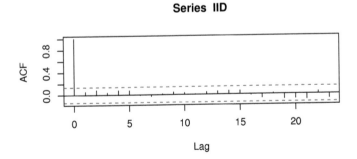

FIGURE 1.5:   A series with IID errors. Below, ACF plot showing autocorrelation of the IID series at a range of lags.

Niche Modeling

```
> par(mfrow = c(2, 1))
> y <- rnorm(200)
> MA <- na.omit(filter(y, filter = rep(1, 10)/10))
> plot(MA, type = "l")
> lines(y, col = "gray")
> acf(MA)
```

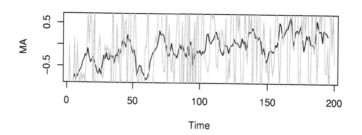

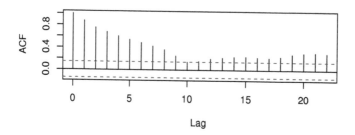

FIGURE 1.6: A moving average of an IID series. Below, the ACF shows oscillation of the autocorrelation of the MA at increasing lags.

The moving average is often called a filter, as it removes elements of the signal, particularly high frequency variations and leaves low frequency ones.

1.4.2.3 Random Walk

A random walk is a series where each value is dependent on the previous, plus some noise. A series with a random walk can be created by taking the cumulative sum of a random IID series. This is done using the *cumsum* function in R. The ACF plot shows autocorrelations in a random walk are extremely persistent, with significant correlation between even over 100 lags (Figure 1.7).

```
> par(mfrow = c(2, 1))
> walk <- ts(cumsum(rnorm(1000)))
> plot(walk, type = "l")
> acf(walk, lag.max = 100)
```

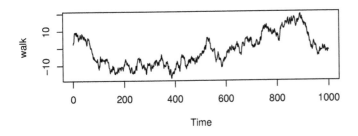

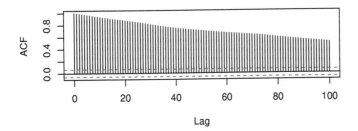

FIGURE 1.7: A random walk from the cumulative sum of an IID series. Below, the ACF plot shows high autocorrelation at long lags.

Series where the average of the numbers is finite tend to stay roughly level no matter how many random numbers are generated. Series where the mean, variance and autocorrelation structure do not change over time are termed *stationary*. A random walk however, is *non-stationary*, as the average value in the infinite limit is infinity.

Non-stationary series like random walks often appear to 'drift' or 'trend', continuing to move up or down for long sequences. These trends can easily be mistaken for a real, causal relationship with other factors. However, the trend is not caused by anything other than the random characteristics of the series and particularly not due to any external factor. In fact, in a random walk the series can diverge arbitrarily far from the starting point. Behavior ranging from the movement of organisms to the evolution of stock prices has been modeled using random walks.

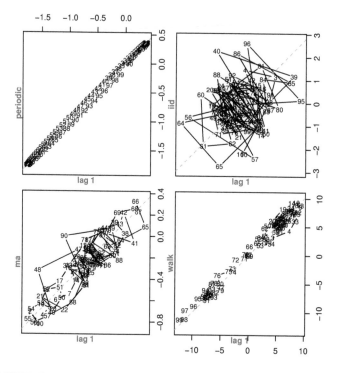

FIGURE 1.8: Lag plots of periodic, random, moving average and random walk series.

Another visual tool for examining the autocorrelation structure of data is the lag plot. A lag plot shows the value of a time series against its successive values (lag 1). The lag plot in Figure 1.8 allows easy discrimination of three main types of series:

- random, shown as a cloud,

- autocorrelated, shown as a diagonal, and

- periodic, shown as circles.

1.4.3   Random fields

A random field is a list of random numbers whose values are mapped onto a space of n dimensions. All of the forms of the previous functions can apply to two dimensions such as the distribution of a species over the landscape, to the spatial distribution of price increases for real estate in a city. The map is

```
> gray <- gray(1:10/10)
> image(matrix(rnorm(10000), 100, 100), col = gray,
+       labels = F)
```

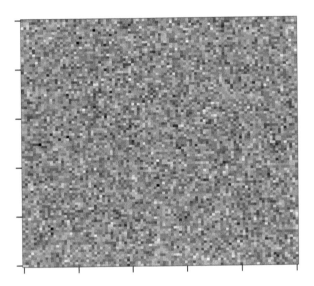

FIGURE 1.9:   An IID random variable in two dimensions.

the most common example of 2D distributions. Random fields simulate the error structure in 2D map data.

To demonstrate a random field in R we generate a two dimensional matrix of random numbers and display it with the image command.

The R random fields library can generate different types of 2D autocorrelation structure. Figure 1.10 shows a field with autocorrelation between neighboring points.

Diagnostics in 1D can usually be applied equally well to 2D data, simply by converting them to a vector. Figure 1.11 shows the ACF for the 2D Gaussian Random Field simulated data. The ACF shows a second peak of correlation at a lag of 100, corresponding to the dimensions of the matrix, when each point becomes correlated with itself. These features are called artifacts and are something to watch out for in any form of analysis.

Niche Modeling

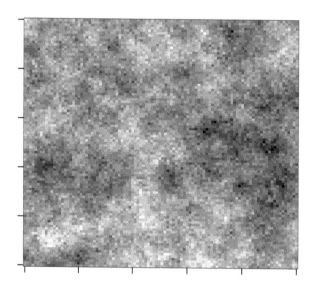

FIGURE 1.10:   An example of a Gaussian field, a two dimensional stochastic variable with autocorrelation.

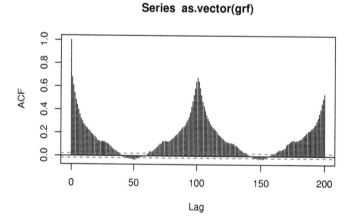

FIGURE 1.11:   The ACF of 2D Gaussian field random variable, treated as a 1D vector.

1.5 Summary

There are many forms of mathematical and statistical relationships used in niche modeling. The main purpose in this chapter has been to introduce these functions, their generation with the statistical language R, and the use of diagnostic tools such as the ACF and lag plots. These functions can be generalized in one or two dimensions.

It would be worthwhile at this point to install R, experimenting with the snippets yourself to become familiar with the various functions. Using with them in this way will improve your intuition for modeling, and most importantly prepare you for recognizing the possible things that can go wrong during analysis.

Chapter 2

Data

Information systems are combinations of computer technologies that help us to store, find, and otherwise handle the information we need. In practice, the central component of an information system is a database. However, a database, by itself, is not enough for niche modeling.

Many people rely on spreadsheets for handling their data. In contrast, for larger datasets and multi-user installations, the conventional approach to information systems is three-tiered, consisting of the following:

- a database,

- an analysis system, and

- a presentation system.

For example, a low-cost system for niche modeling might use a database, the statistics language R for analytics, and a browser for presentation of results.

While there are advantages in partitioning operations in this way, inefficiencies and errors are introduced when transferring data between applications, particularly in passing data in and out of databases. In keeping with the main theme of this book – successful niche modeling and recognition of sources of error – eliminating a possible source of errors is one of the advantages of a single system like a spreadsheet.

Fully integrated systems can also be very efficient. A novel approach developed for the financial industry integrates database and analysis functions into the one, memory resident application. The vector-based K language has produced very fast trading applications – over 100 times faster than equivalent three-tiered applications [Whi04].

In this section we demonstrate the use of R as a database. While not recommended for large datasets, it may be possible to dispense with an external database by replicating relational database operations including *select* and *join* in R. As well as simplicity and efficiency for smaller systems, a simple database example helps to build knowledge of the R's powerful indexing operations.

SQL or structured query language is one of the main languages used in databases. While not going into the syntax of this language, we replicate expressions in SQL with operations written in R.

2.1 Creating

Data must be moved in and out of a database. Where one might use *import* or *mysqlimport* in MySQL, here we read a small local file of locations where a species was sighted that might have been saved in a spreadsheet. The operation *read.table* also permits reading from a URI, so data can be read directly from the web.

```
> locations <- read.table("obs.txt", header = TRUE)
```

TABLE 2.1:   Example
data consisting of field
observations with locations.

	id	X	Y
11	ML240	110.00	111.00
2	ML240	211.00	102.00
3	ML240	123.40	114.30

A database is composed of a set of tables. The R data structure called a *dataframe* serves the purpose of tables in a relational database. A data frame is really a list of vectors (columns) of equal length, ideal for storing all kinds of information. To simulate a relational database table, the data frame name serves as the table name, and column names serve as the table attributes. Comprehensive information about a data frame, or any R object, can be listed with the command *attributes*.

```
> attributes(locations)

$names
[1] "id" "X"  "Y"

$class
[1] "data.frame"
```

```
$row.names
[1] "11" "2"   "3"
```

2.2   Entering data

Manipulating and entering data in R is perhaps easier than most databases. To insert rows at the bottom of a table we use *rbind*. To insert a column, say of number of animals seen at each observation, we would use *cbind*.

```
> locations <- rbind(locations, data.frame(id = "PC101",
+     X = 113.4, Y = 114.3))
> locations <- cbind(locations, Number = c(2, 1, 3,
+     1))
> locations

      id     X     Y Number
11 ML240 110.0 111.0      2
2  ML240 211.0 102.0      1
3  ML240 123.4 114.3      3
1  PC101 113.4 114.3      1
```

Altering tables is easy in R. Whereas a table is modified in a relational database with the *drop* command, in R a column is deleted by assignment of NULL. Assignment also changes the name of a table, and the old *locations* object can be deleted with *rm*.

R automatically assigns names to rows but not the ones we want. A unique index for each observation could be added to the table as row names as shown below. Changing the names of columns is also easy in R, as shown.

```
> row.names(locations) <- c("12", "13", "14", "15")
> obs <- locations
> rm(locations)
> names(obs) <- c("id", "X", "Y", "n")
> obs

      id     X     Y n
12 ML240 110.0 111.0 2
13 ML240 211.0 102.0 1
14 ML240 123.4 114.3 3
15 PC101 113.4 114.3 1
```

2.3   Queries

Of course the previous operations are preliminaries relative to the main purpose of a relational database, which is to perform queries. As before, R can easily be coerced to replicate SQL statements for queries. The following are examples of how concisely the indexing in R can mimic the corresponding SQL select statement.

2.3.0.1   SQL: SELECT * FROM obs

```
> obs

      id    X      Y n
12 ML240 110.0 111.0 2
13 ML240 211.0 102.0 1
14 ML240 123.4 114.3 3
15 PC101 113.4 114.3 1
```

2.3.0.2   SQL: SELECT * FROM obs WHERE rownum == 2 or rownum ==3

```
> obs[2:3, ]

      id    X      Y n
13 ML240 211.0 102.0 1
14 ML240 123.4 114.3 3

> obs[c(FALSE, TRUE, TRUE, FALSE), ]

      id    X      Y n
13 ML240 211.0 102.0 1
14 ML240 123.4 114.3 3
```

Operations on columns have a shorthand form. The dollar selects a column from a table by name.

2.3.0.3   SQL: SELECT id FROM obs

```
> obs$id

[1] ML240 ML240 ML240 PC101
Levels: ML240 PC101
```

A more flexible form of selection with an arbitrary list of column names as an argument is *subset*.

2.3.0.4 SQL: SELECT X,Y FROM obs

```
> subset(obs, select = c("X", "Y"))

      X     Y
12 110.0 111.0
13 211.0 102.0
14 123.4 114.3
15 113.4 114.3

> subset(obs, select = c(2, 3))

      X     Y
12 110.0 111.0
13 211.0 102.0
14 123.4 114.3
15 113.4 114.3
```

The following example illustrates the combination of row and column selection.

2.3.0.5 SQL: SELECT * FROM obs WHERE id = PC101

```
> obs[obs$id == "PC101", ]

      id     X     Y n
15 PC101 113.4 114.3 1

> subset(obs, id == "PC101")

      id     X     Y n
15 PC101 113.4 114.3 1
```

Another important feature of database languages is functions for aggregating data, such as summing and counting. Say we wanted to find out how many animals of each species had been observed. In SQL we would use a combination of select and the count or sum functions to perform aggregation.

The following are examples of R implementations of typical SQL queries using the *dim* and *aggregate* functions. The first counts the number of unique observations. The second sums the number of animals of each species observed at a site.

2.3.0.6 SQL: SELECT id, count(*) FROM obs

```
> dim(obs[obs$id == "PC101"])[1]
```

```
[1] 4
```

2.3.0.7 SQL: SELECT * FROM obs GROUP BY id

```
> aggregate(obs$n, list(obs$id), FUN = sum)
```

```
  Group.1 x
1   ML240 6
2   PC101 1
```

2.4 Joins

Equally important as select operations on relational databases are joins. A join is the Cartesian product of two tables formed on a common index column.

Say we wanted to list the animals of each species by their common name instead of the field code. To demonstrate this we create another table for our species database called species, containing the common names for the species of interest. The *merge* operation together with *subset* obtains the common names of species at each location, thus joining the two tables. In the second case, the subset operation produces a list of locations where the Puma has been observed.

2.4.0.8 SQL: SELECT * FROM species, obs WHERE obs.id = species.id

```
> species <- data.frame(id = c("PC101", "ML240", "J2"),
+     Name = c("Puma", "Mountain Lion", "Jaguar"))
> merge(species, obs, by.species = "id", by.obs = "id")
```

```
     id          Name     X     Y n
1 ML240 Mountain Lion 110.0 111.0 2
2 ML240 Mountain Lion 211.0 102.0 1
3 ML240 Mountain Lion 123.4 114.3 3
4 PC101          Puma 113.4 114.3 1
```

2.4.0.9 SQL: SELECT X, Y FROM Obs, Species WHERE Obs.id = Species.id AND Species.name=Puma;

```
> subset(merge(species, obs), Name == "Puma", select = c(X,
+     Y))

    X     Y
4 113.4 114.3
```

2.5 Loading and saving a database

Finally, there are commands for loading and saving a database. R allows all forms of information to be loaded or saved as distinct objects. The command *load* brings a saved R object into the current environment. The command *save* writes R objects to file.

The role of database can be played by files, or objects. In addition, all objects in the user environment, the set of all data and functions defined by the user, can be collectively stored. This is easier than dealing with separate objects. The command *ls* or *object* lists the data sets and functions defined currently by the user, while *save.image* saves all user-defined objects in the current environment.

```
> save.image(file = "test.Rdata")
> load("test.Rdata")
```

2.6 Summary

This chapter has demonstrated how to manage data in R using data frames as tables, including replicating the basic spreadsheet and relational database operations. While a database is necessary for large-scale data management, it is clear that R can serve the basic functions of a database system. This can be an advantage, as data do not need to be moved between systems, eliminating conversion problems and increasing the numbers of errors.

Chapter 3

Spatial

Niche modeling often uses spatial information about the distribution of species, so Geographic Information Systems (GIS) are often the tool of choice when faced with managing and presenting large amounts of geographic information.

However, when the main interest is analysis of moderate amounts of data, as in the previous chapter when R was used as a relational database, R can be used to perform simple spatial tasks. This both avoids the need for a separate GIS system when not necessary, and helps to build knowledge of advanced use of the R language.

On the down side, native R operations are not very efficient for spatial operations on large matrices. Vectors and matrices in R, the form necessary for mathematical operations, use quite a lot of memory, thus limiting the size of the data that can be handled. Another approach is to perform basic niche modeling functions on large sets of data with image processing modules. Examples of the image processing package called *netpbm* performing fundamental analytical operations for modeling are given.

3.1 Data types

The two main types of data representing geographic information are:

- surface features, such as temperature and rainfall, called raster, and

- linear features, such as roads, streams or areas represented by polygons.

A raster is a regular grid of numbers, where each number represents the value of a variable in a regular area, or cell. A raster can be represented in R either as a matrix or a vector. The contents of the cells can be integers, floating point numbers, characters or raw bytes.

In Figures 3.1 and 3.2 are examples of two ways we might generate a raster to use in analysis: by simulation or input from a data file. The first is a

Niche Modeling

```
> palette(gray(seq(0, 0.9, len = 30)))
> pts <- seq(-pi, pi, by = 0.1)
> z <- sin(pts) + (0+1i) * cos(pts)
> points <- cbind(Re(z) * 20 + 62, Im(z) * 10 + 26)
> t1 <- matrix(255, 124, 52)
> t1[points] <- 0
> image(t1, col = 1:30, labels = F)
```

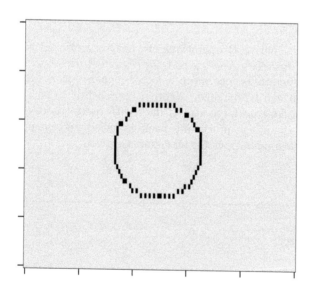

FIGURE 3.1:   Example of a simple raster to use for testing algorithms.

```
> x <- readBin("../ZM/layer06", what = "raw", n = 124 *
+     52)
> image(matrix(as.numeric(x), 124, 52), ylim = c(1,
+     0), col = 1:30, labels = F)
```

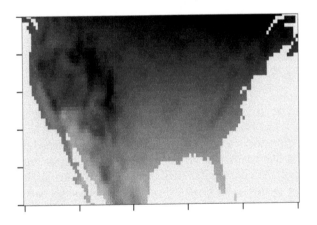

FIGURE 3.2: Example of a raster from an image file representing the average annual temperature in the continental USA.

matrix generated with a sequence of numbers and displayed with the image command. The second is a raw raster dataset stored as an image. Simulated datasets such as these are essential for verifying that code is working correctly, and should always be used to test algorithms before testing on real data.

The format of the image data is often a gray scale image, where the data in each cell, called a pixel in an image, is a single byte. Images are an efficient and convenient format of storing large amounts of data, although they have the disadvantage of not containing geographic coordinates to enable alignment with other maps and geographic data.

Below, an example of the portable gray map or pgm format used in netpbm has many similarities to most image formats. The code, 'P2', called a magic number that identifies the type of the file and next line contains optional comments. The dimensions of the image follow, then the number of colors in the image, and finally the data in each pixel. The disadvantage of the pgm format is the less efficient format without compression of the image data, and the limitation of the range of values limited from 0 to 255. The advantage of the format is simplicity and ease of use.

```
P2
# feep.pgm
24 7
15
0  0  0  0  0  0  0  0 ...
```

The second main type of data are point locations. Sets of related points can be used to form lines like roads and streams, or polygon shapes representing areas. The ordering of the coordinates defines the connections in the road or shape. These can be represented in R either as a matrix with two columns, one for each x and y coordinate, or as vector of complex numbers.

The different forms of data, rasters and points combined together make maps. While basic R operations do not support mapping operations, contributed additional modules or extensions do. There are, however, a number of other useful features in the base R distribution for simple presentation graphics. Figure 3.4 also illustrates R's contour function applied to the previous matrix.

3.2 Operations

While many operations in professional GIS's produce professional looking maps, basic R is devoted to analytical operations. Some mathematical oper-

```
> plot(z, type = "l", labels = F)
> points(rnorm(100)/3, rnorm(100)/3, cex = abs(rnorm(100)))
```

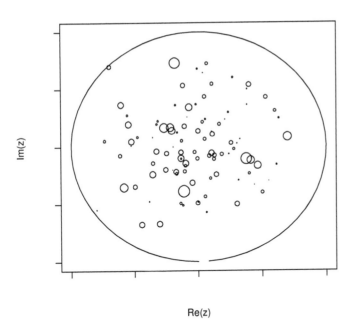

FIGURE 3.3: Examples of vector data, a circle and points of various sizes.

```
> contour(1:124, 1:52, matrix(as.numeric(x), 124,
+       52), ylim = c(51, 1))
```

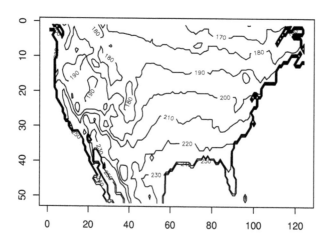

FIGURE 3.4:   A contour plot generated from the annual temperature raster map.

```
> par(mfcol = c(1, 2))
> values <- as.numeric(t1)
> image(t1, col = 1:30, labels = F)
> hist(values, main = "")
```

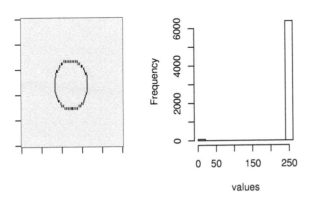

FIGURE 3.5: Simulated image with distribution of values shown in a histogram.

ations can be applied to matrices, and vectors to help to prepare spatial data to answer statistical questions follow.

3.2.1 Rasterizing

The first operation is rasterizing – plotting points onto a raster. This is very easy using the indexing operations in R as used to generate the circle in Figure 3.1.

Histograms are helpful for examining the distributions of environmental values. In the histogram of classes in Figure 3.5 a large proportion of values are 255. This is due to using the value of 255 to represent areas other than the featured shape.

3.2.2 Overlay

Overlay of values is a common operation in ecological niche modeling. Arithmetic operations can modify values in another vector the way we require. Figure 3.6 is an example of overlay by multiplication. Masking, as shown in chapter one, is a form of overlay.

```
> par(mfcol = c(1, 2))
> t1[t1 == 255] <- 1
> comb <- t1 * as.numeric(matrix(x, 124, 52))
> image(comb, col = 1:30, ylim = c(1, 0))
> values <- as.numeric(comb)
> hist(values, main = "")
```

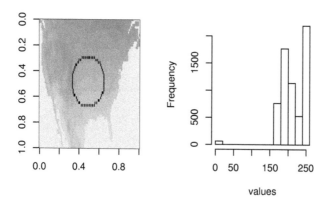

FIGURE 3.6:   Application of an overlay by multiplication of vectors. The resulting distribution of values is shown in a histogram.

 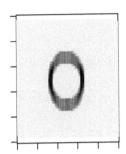

FIGURE 3.7: Smoothing of simulated image, first in the x direction, then in the y direction.

Arithmetic operations in the netpbm image processing module also produce overlay or masking. Many options for combining images arithmetically are available:

```
>pamarith -add | -subtract | -multiply | -divide | -difference |
-minimum | -maximum | -mean | -compare | -and | -or |
-nand | -nor | -xor | -shiftleft | -shiftright pamfile1 pamfile2
```

For example, a masking operation could make use of the limited range of the single byte in each cell with the following command. If the areas to be masked have value 255 (white) then all areas in the image to be masked, in back.ppm, will have the value 255 after the operation.

```
>pamarith -add back.ppm mask.ppm > masked.ppm
```

3.2.3 Proximity

Proximity is often an important relationship to capture in ecological analysis. Convolution is the operation whereby we use neighboring values to determine the value of the central cell. Filters in R achieve convolution in one direction. The transpose of the matrix and filter application completes the smoothing in the other direction (Figure 3.7).

Image processing packages usually have a smoothing operation that achieves the same purpose. The convolution program in neppbm is *pnmsmooth* or *pnmconvol.* The utility *pnmconvol* uses a convolution matrix file to specify the exact profile the smoothing over the neighborhood of the image. The parameters emphwidth and emphheight specify the dimensions of the convolution matrix.

```
>pnmsmooth [-width=cols] [-height=rows] [-dump=dumpfile] [pnmfile]
```

```
>pnmconvol convolution_matrix_file [-nooffset] [pnmfile]
```

```
    P2
    3 3
    18
    2 2 2
    2 2 2
    2 2 2
```

3.2.4 Cropping

Cropping is the trimming of unwanted parts of a 2D matrix to leave the parts necessary for analysis. For example, a continental map may be cropped to remove the area surrounding the set of points where a species occurs, to develop more specific regional models.

This is done by *pamcut* in netpbm as follows:

```
>pamcut [-left colnum] [-right colnum] [-top rownum]
[-bottom rownum] [-width cols] [-height rows] [-pad] [-verbose]
[left top width height] [pnmfile]
```

In R one can develop an array of indices for each of the cells in the rectangle required with a command called *expand.grid*. This is however not suitable for the large data matrices typically used in spatial analysis for niche modeling.

3.2.5 Palette swapping

Predictive models can involve very complex data-mining and other statistical techniques. Palette swapping is an efficient way of predicting using images reducing the prediction process to its bare essentials.

All models are essentially generalizations, or simplifications, of more specific information, such as observations. The type of generalization expresses the theory in mathematical terms. As such, one of the simplest forms of generalization is categorization, or clustering, where a large number of dissimilar items are sorted into a smaller number of bins, based on their similarity. Given a set of bins or categories developed from a set of examples, new items can be categorized into these bins. In this way, a categorization, or clustering, can serve as a predictive model.

In R the basic operation for clustering is called *kmeans*. In *kmeans*, the data to be clustered are partitioned into k groups in order to minimize the sum of squares from points to the assigned cluster center. Cluster centers are typically at the mean of the set of all data points in the same category.

Color quantization or color reduction is a similar operation in image processing. Reducing the number of colors will compress the size of an image at the expense of the number of colors. The utility in netpbm for this purpose is called *ppmquant*.

```
>ppmquant colors pnmfile
```

3.2.5.1 Prediction

Index mapping is the fundamental operation of mapping one set of values into another set of values. Prediction using palette swapping can be seen as a mapping of one set of palette indices into another.

A very simple and efficient prediction system treats colors as indices of environmental values. To use with a model, index mapping changes the values in cells from their original value, to a probability for each specific class.

In image processing, index mapping is called palette mapping, a process where the original colors are substituted for a new set. Palette swapping essentially maps one set of colors into another set.

The main limitation of index mapping is the characterization of continuous environmental values as categories. When implemented as palette swapping on images, there are limits on the number of dependent variables – one for each of the red, green and blue channels.

To create a simple niche model prediction with palette swapping, we take a three-color image, with each of red, green and blue representing values of environmental variables, and map it into a gray scale image, with intensity representing the probability of occurrence of the species in each (environmental) category of color.

In practice, a palette file defines the mapping. The utility for this purpose in the netpbm package is *pamlookup*, invoked with an image as a lookup table for mapping the old into the new palette.

```
>pamlookup -lookupfile=lookupfile -missingcolor=color
[-fit] indexfile
```

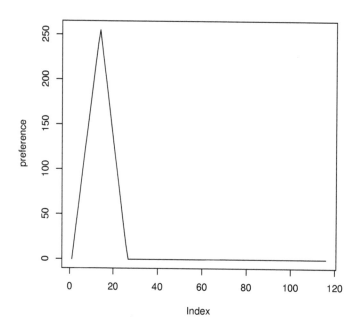

FIGURE 3.8: A hypothetical niche model of preference for crime given environmental values.

Note this is a very efficient operation as the data in the image do not change, only the small set of values in the palette of the image.

For example, say we have an image that represents a pattern of environmental values. For concreteness, the donut in Figure 3.9 could be the vicinity of a ring-road, the edges of an urban area, or a similar feature. Say we predict that certain values of a variable are of interest, due to past crimes. The frequency of those crimes with respect to the environmental values constitutes a niche model, as shown by the peaked distribution over the environmental values in Figure 3.8.

Swapping the colors in the original image with the new colors given by the function in Figure 3.8 (i.e. mapping the values on the x-axis to the values on the y-axis) changes the double convoluted circle in Figure 3.7 into Figure 3.9, the darker areas at the edges of the shape representing a prediction of the regions of high probability of crime.

```
> par(mfcol = c(1, 1))
> f1 <- as.vector(t(f2))
> prob = seq(255, 1, -1)
> result <- prob[as.integer(f1 * 255 + 1)]
> image(matrix(result, 124, 52), col = 1:30, labels = F)
```

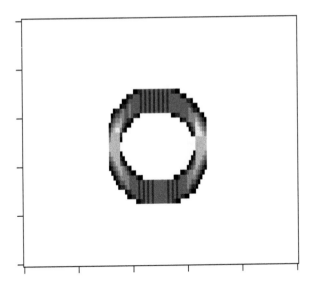

FIGURE 3.9: The hypothetical prediction of probability of crime, after palette swapping.

3.3   Summary

R and image processing operations can perform the elementary operations necessary for niche modeling. While these do not replace a GIS for many features, the basic arithmetic concepts of spatial analysis can be implemented using simple tools.

All analysis represents a theoretical approximation of a more complex real-world concept. As such, any approach has strengths and weaknesses. There is often a trade-off between theoretical correctness and computational efficiency. It is important to be aware of these trade-offs when performing analysis, and to try to understand and verify the correctness of any heuristic tool or procedure for the specific application at hand.

Chapter 4

Topology

The focus of topology here is the study of the subset structure of sets in the mathematical spaces. Topology can be used to describe and relate the different spaces used in niche modeling. A topology is a natural internal structure, precisely defining the entire group of subsets produced by standard operations of union and intersection. Of particular importance are those subsets, referred to as *open sets*, where every element has a neighborhood also in the set. More than one topology in X may be possible for a given set X.

Examples of subsets in niche modeling that could form topologies are the geographic areas potentially occupied by a species, regions in environmental space, groups of species, and so on.

Application of topological set theory helps to identify the basic assumptions underlying niche modeling, and the relationships and constraints between these assumptions. The chapter shows the standard definition of the niche as environmental envelopes around all ecologically relevant variables is equivalent to a box topology. A proof is offered that the Hutchinsonian environmental envelope definition of a niche when extended to large or infinite dimensions of environmental variables loses desirable topological properties. This argues for the necessity of careful selection of a small set of environmental variables.

4.1 Formalism

The three main entities in niche modeling are:

S: the *species*,

N: the *niche* of environment variables, and

B: geographic space, where the environmental variables are defined.

The relationships between these entities constitute whole fields of study in themselves. Most applications of niche modeling fall into one of the categories in Table 4.1.

TABLE 4.1:   Links between geographic, environmental and species spaces.

	S	N	B
S	interspecies relationships	–	–
N	habitat suitability	correlations	–
B	range predictions	geographic information	autocorrelation

Niche modeling operates on the collection of sets within these spaces. That is, a set of individuals collectively termed a species, occupies a set of grid cells, collectively termed its range, of similar environmental conditions, termed its niche. Thus a niche model $\mathcal{N}$ is a triple:

$$\mathcal{N} = (S, N, B)$$

The niche model is a general notion applicable to many phenomena. Here are three examples:

- Biological species: e.g. the mountain lion *Puma concolor*, the environment variables might be temperature and rainfall, and space longitude and latitude.

- Consumer products: e.g. a model of digital camera, say the Nikon D50, environment variables for a D50 might be annual income and years of photographic experience, and space the identities of individual consumers.

- Economic event: e.g. a phenomenon such as median home price increases greater than 20%, the variables relevant to home price increases would be proximity to coast, family income, and the space of the metropolitan areas.

A niche model can vary in dimension. Here are some examples of dimensions of the geographic space B:

- zero dimensional such as a set, e.g. survey sites or individual people,

- one dimensional such as time, e.g. change in temperature or populations,

- two dimensional such as a spatial area, e.g. range of a species,

- three dimensional such as change in range over time.

While examples of contemporary niche modeling can be seen in each of these dimensions, many examples in this book are one dimensional, particularly in describing the factors that introduce uncertainty into models, because a simpler space is easier to visualize, analyze and comprehend. All results should extend to studies in higher dimensions.

Dimensions of environmental space N, in Chapter 4, concern the implications of extending finite dimensional niche concepts into infinite dimensions.

Dimensions of species, one species for each dimension, relates to the field of community ecology through inter-specific relationships.

Here we restrict examples to one species, and one S dimension.

4.2 Topology

There are a number of other ways to describe niche modeling. There are a rich diversity of methods to predict species' distribution and they could be listed and described. Alternatively, biological relationships between species and the environment could be emphasized, and approaches from population dynamics used as a starting point. While useful, these are not the approaches taken in this book, preferring to adhere to examination of fundamental principles behind niche modeling.

Topology is concerned with the study of qualitative properties of geometric structures. One of the ways to address the question – What is niche modeling? – is to study its topological properties.

4.3 Hutchinsonian niche

Historically, the quantitative basis of niche modeling lies in the Hutchinsonian definition of a niche [Hut58]. Here that set of environmental characteristics where a species is capable of surviving was described as a 'hypervolume' of an n-dimensional shape in n environmental variables. This is a generalization of more easily visualizable lower dimensional volumes, i.e.:

- one, an unbroken interval on the axis of an environmental variable, representing the environmental limits of survival of the species,

- two, a rectangle,

- three, a box,

- n dimensions, hypervolumes.

This formulation of the niche has been very influential, in part because in contrast to more informal definitions of the niche, it is easily operationalized by simply defining the limits of observations of the species along the axes of a chosen set of ecological factors.

4.3.1   Species space

Hutchinson denotes a species as S_1 so the set of species is therefore denoted S. In its simplest form the values of the species S_1 are a two valued set, presence or absence:

$$S_1 = \{0, 1\}$$

Alternatively the presence of a species could be defined by probability:

$$S_1 = \{p | p \in [0, 1]\}$$

4.3.2   Environmental space

Using the notation of Hutchinson the niche is defined by the limiting values on independent environmental variables such as x_1 and x_2. The notation used for the limiting values are x_1', x_1'' and x_2', x_2'' for x_1 and x_2 respectively. The area defined by these values corresponds to a possible environmental state permitting the species to exist indefinitely.

Extending this definition into more dimensions, the fundamental niche of species S_1 is described as the volume defined by the n variables $x_1, x_2, ..., x_n$ when n are all ecological factors relative to S_1. This is called an n-dimensional hypervolume N_1.

4.3.3 Topological generalizations

The notion Hutchinson had in mind is possibly the Cartesian product. If sets in environmental variables x_i are defined as sets of spaces X_i, then N_1 is a subset of the Cartesian product X of the set $X_1, ..., X_n$, denoted by

$$X = X_1 \times ... \times X_n, \text{ or}$$

$$X = \prod_{i=1}^{n} X_i$$

In a Cartesian product denoted by set X, a point in an environmental region is an n-tuple denoted $(x_1, ...x_n)$.

The environmental region related to a species S_1 is some subset of the entire Cartesian space of variables X. The collection of sets has the form

$$\prod_{i \in J} X_i$$

Setting a potentially infinite number $i \in J$ to index the sets, rather than a finite i equals 1 to n is a slight generalization. The construct captures the idea that the space X_i could consist of an infinite number of intervals. This generalizes the n-dimensional hypervolume for a given species in S, so that the space may encompass a finite or infinite number of variables.

Another generalization is to define each environmental variable x_i as a topological space. A topological space T provides simple mathematical properties on a collection of open subsets of the variable such that the empty set and the whole set are in T, and the union and the intersection of all subsets are in T. The set of open intervals:

$$(x_i', x_i'') \text{ where } x_i', x_i'' \in \mathbb{R}$$

is a topological space, called the standard topology on $\mathbb{R}$.

Where each of the spaces in X_i is a topology, this generates a topology called a box topology, describing the box-like shape created by the intervals. An element of the box topology is possibly what Hutchinson described as the the n-dimensional hypervolume N_1 defining a niche.

4.3.4 Geographic space

There are differences between the environmental space N and the geographical space B. While the distribution of a species may be scattered over many discrete points in B, the shape of the distribution in N should be fairly compact, representing the tendency of a species to be limited to a fairly small environmental region. Perhaps the relevant concept from topology to describe this characteristic is *connected*. When the space N is connected, there is an unbroken path between any two points. However, the same is not true of the physical space B where populations could be isolated from each other.

4.3.5   Relationships

There is a particular type of relationship between N and B. Every species with a non-empty range should produce a non-empty niche in the environmental variables. Moreover, a single point in the niche space N will have multiple locations in the geographic space B, but not vice versa.

The relationship of niche to geography is a function. A function f is a rule of assignment, a subset r of the Cartesian product of two sets $B \times N$, such that each element of B appears as the first coordinate of at most one ordered pair in r. In other words, f is a function, or a mapping from B to N if every point in B produces a unique point in N:

$$f : B \longrightarrow N$$

The inverse is not true, as a point in N can produce multiple points in B, those geographic points with the same niche, due to identical environmental values.

One generalization used extensively in machine learning is to assume a set of real-valued functions $f_1, ..., f_n$ on B known as *features* such as the variable itself, the square, the product of two features, thresholds and *binary features* for categorical environmental variables [PAS06]. A binary feature takes a value of 1 wherever the variable equals a specific categorical value, and 0 otherwise.

In another functional relationship g from N to S, each species occupies multiple niche locations, but one niche location has a distinct value for the species space S, such as a probability.

$$g : N \longrightarrow S$$

Similarly, there is a functional relationship h from B to S where each species may occupy multiple geographic points, but there is a unique value of a species at each point.

$$h : B \longrightarrow S$$

The natural mappings h from physical range B to the species S are referred to as the *observations*. An alternative mapping, from B via the niche N to S, is referred to as the *prediction* of the model. The similarity between these mappings is the basis of assessments of accuracy.

$$g(f(B)) \sim h(B).$$

4.4 Environmental envelope

We now consider how to operationalize these theoretical set definitions.

The approach of defining limits for each of the environmental variables captures the sense of a niche as understood by ecologists: that the occurrence of species should be limited by a range of environmental factors, and that an envelope around those ranges would have predictive utility. This approach was used in environmental envelopes, one of the first niche modeling tools first used in an early study of the distribution of snakes in Australia by Henry Nix [Nix86].

However, the approach has some practical problems.

4.4.1 Relevant variables

The Hutchinsonian definition suggests that the box continues in n-dimensions until all ecological factors relevant to S_i have been considered [Hut58].

There are a number of problems with this definition. One problem stems from the vagueness of what is meant by an ecologically relevant factor. The formalism provides no way to weight variables by importance, or exclude variables from the niche. Another problem is the number of potentially relevant ecological factors is unlimited.

4.4.2 Tails of the distribution

The environmental envelope defines limits for the species largely by the tails of the probability distribution. The tails of a probability distribution usually have the smallest probabilities, the least numbers of samples, and hence estimated with the least certainty. Hence a definition based on limits must be statistically uncertain, or at least less certain than a range that was defined, say, via a type of confidence limits using mean values and variance.

Often to reduce the variability of the range limits the niche includes only the 95% percentile of locations from B. Unfortunately this approach produces a progressive reduction in ecological area with each variable, leading to underestimation of species' potential ranges [BHP05]. Niche descriptions such as based on Mahalanobis distances allow more flexible descriptions of the distribution and have been shown to be more accurate [FK03].

4.4.3   Independence

The box-like shape only applies to independent variables, but species rarely fit within a sharp box-like shape. Niche descriptions based on more flexible descriptions of the shape of the space do not make such strong assumptions as independence between variables [CGW93].

4.5   Probability distribution

While the above approaches to correcting the deficiencies of environmental envelopes led to some improvements, an essential component was missing in S. In the Hutchinsonian niche, the environmental envelope of a species can only take values of 1 or 0. Environmental envelopes do not explicitly estimate probability. That is, while they define a region in space, the variation in probability within that region is undefined. Thus what is required to define a niche is more like the notion of a probability density.

$$P(x \in N) = \int_N P(x)dx$$

A probability distribution, more properly called a probability density, assigns to every interval of the real numbers a probability, so that the probability axioms are satisfied. The probability axioms are the natural properties of probability: values defined on a set of events are greater than zero, that the probability of all events sum to one, and that the union of independent events is the sum of the individual probabilities of the events.

In technical terms, expressing a niche in this way requires the extension of the simple Hutchinsonian definition of a niche to a theoretical construct called a *measure*. A measure is a function that assigns a number, e.g., a 'size', 'volume', or 'probability', to subsets of a given set such that it is possible to carry out integration.

With a niche defined as a probability distribution the probability at each point E in the environmental space N satisfies axioms of a measure:

$$Pr[0] = 0$$

and countable additivity

$$Pr(\bigcup_{i=1}^{\infty} E_i) = \sum_{i=1}^{\infty} Pr(E_i)$$

This is not true of the physical space B. Each distinct point may have a probability, as a result of the mapping defined previously, that could be used in the sense of a probability of species occurrence or habitat suitability. However, the sum of the probabilities over all points in physical space is not less than one, so this is not a probability distribution.

So the more general approach to niche modeling, an extension of the Hutchinsonian niche, is the statistical idea of the probability distribution. Here the niche model is a probability distribution over the environmental variables.

This definition of the niche as a probability distribution has some important implications. Based on this definition, the 'entity' being modeled is probabilistic, not an actual physical object that exists or not, and not a quantity such as population density of animals or group of plants. Probabilistic definitions are suitable for expressing fairly vague concepts, such as preference of habitat suitability. In a way the object of the niche modeling is similar to a quantum entity – in the realm of possibility rather than actuality.

Such a viewpoint is useful if one is careful not to carry the metaphor too far, partly because the fundamental constraints that govern microscopic physical systems, such as conservation of energy laws, do not hold.

4.5.1 Dynamics

Niche models are sometimes called equilibrium models, as generally the niche represents a stable relationship of a species to its environment. Stability in this sense refers to the overall stability of a population despite non-equilibrium disturbances such as annual cycles and episodic threats. For example, the processes that lead to expansion of the range of the species balance the processes that lead to contraction and result in an equilibrium.

But equilibrium assumptions are not necessary to develop these models. Any form of reasonably 'stable' probability distribution can produce a dynamic distribution. For example, while migrating species move in relation to their environment, it has been shown that many are 'niche followers' by remaining in a fairly constant climate as the seasons change [JS00]. Invasive species are another example of species not at 'equilibrium' but generally only spreading to similar environmental niches to those occupied in their host country [Pet03].

That is, the assumptions of equilibrium are for the space N and should not be confused with equilibrium, or stability, in the geographic space B.

4.5.2   Generalized linear models

Given the probability structure for a niche we need to define a way of operationalizing the concept for prediction. Perhaps the most familiar approach is to define the probability over the sums of environmental variables. This is called a logistic regression and are among the most well studied and understood statistical methodologies. In a logistic regression, with probability p of a binary event Y, such as the occurrence or absence of a species, i.e. $p = Pr(S_i = \{1, 0\})$, there is a logit link function between that probability $p \in S$ and the values of the environmental variables $(x_1, ..., x_n) \in N$

$$logit(p) = ln(\tfrac{p}{1-p}) = \alpha + \beta_1 x_1 + \beta_2 x_1^2 + ... + \beta_{2n} x_n^2 = y$$

The expression admits estimation of the parameters $\beta_1, ..., \beta_{2n}$ for the simple linear equation y using least squares regression, i.e. calibrating the model. With the expression below we can calculate p, given y, and thus apply the model $g : N \longrightarrow S$ where (Figure 4.1)

$$p = g(x) = \tfrac{e^y}{1+e^y}$$

4.5.2.1   Naughty noughts

The introduction of statistical rigor helps identify and define problems. An example of one such problem is called the 'naughty noughts', referring to the great many areas with essentially zero probability beyond the range of the species. These include oceans for a terrestrial species, and land for a marine species. Logistic models will be distorted by these and give predictions of positive probability where the species is known to be absent [AM96].

Most well known and used probability distributions, such as the Gaussian distribution, are continuous with finite (though sometimes very small) probability over the whole range. Using these distributions leads to predictions of non-zero probability in obviously inappropriate places.

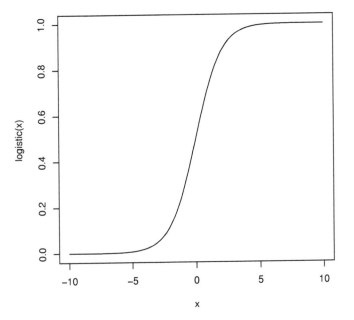

FIGURE 4.1: The logistic function transforms values of y from $-\infty$ to ∞ to the range $[0,1]$ and so can be used to represent linear response as a probability.

The need to eliminate the noughts, by restricting the data over the suitable range, led to the use of truncated distributions, and more flexible ways of defining probability distribution such as Generalized Additive Modelling (GAM).

4.5.2.2   Form of distribution

Actually, the problem of finding the best shape for the probability distribution of a species in N is not trivial. It cannot be taken for granted that a simple linear additive model of environmental variables will be appropriate. Species distributions are not necessarily 'normal' and can be skew, bimodal, exponential or sigmoidal.

They can also potentially have more unusual distributions. While it is generally believed that systems in equilibrium display approximately exponential decay of distributions and correlations, systems far from equilibrium, including climatic, hydrological and biological systems, display power law decay in both distribution and autocorrelations [LL58].

In power law distributions, extreme events to be more likely than expected, known as 'fat tails' or 'long tails'. In these situations, the use of normal curves, which decay exponentially, will tend to greatly underestimate the probability of extreme events. For example, with respect to species dispersal capabilities, empirical studies show that seed dispersal curves decay more slowly than exponential for many, if not the majority of species [PW93]. Both justifying the shape of the distribution and modeling with the range of possible distributions involves difficult and challenging statistical tests using classical statistical approaches.

4.5.2.3   Categorical variables

Another difficulty associated with logistic regression is the treatment of categorical variables such as vegetation types, ecological regions, and so on. In the formalism used in environmental envelopes, the event set on which the niche was defined was a range of variables, with a lower and upper limit. This idea is captured in the continuous probability distribution, defined on all a and b.

$$Pr(a < x < b) = \int_a^b f(x)dx$$

Intuitively the distribution of this probability function varies smoothly, whereas a categorical variable varies by unrelated 'jumps' visualized as bars on a bar graph. In a regression model with continuous variables, the discrete variables in the niche space are no longer connected using a somewhat stronger property of sets in ecological and physical space – one related to topological connectedness [Mun75]. In a connected space X, there exists no separation of N into disjoint non-empty sets whose union is N. Connectedness captures the notion of an entity that is not broken into parts, and gives the niche a sense of wholeness. In another sense, there is always an unbroken path between any two points in the niche of a species.

The categorical variables are usually converted into a set of binary variables for analysis by logistic regression. However, with more categories and more variables the number of variables that would need to be introduced can be prohibitive. For example, if a variable has 100 categories, this procedure would produce 100 new binary variables.

4.6 Machine learning methods

Due in part to the popularity of artificial intelligence, machine learning was applied to the problems posed by niche modeling. Machine learning methods, characterized as heuristic search methods, have been used in a variety of problems where there were no exact analytical solutions.

The popular early methods: decision trees, neural nets and genetic algorithms are loosely based on human cognitive or biological approaches to optimization. In the case of genetic algorithms, the idea is to copy the strategy of biological evolution, to generate a population of models and then iteratively test and refine them until a stable solution is achieved, letting the best rules reproduce, flourish and eventually dominate the population. The GARP approach was an attempt to meld the three traditional approaches in a genetic algorithm that evolved a set of solutions consisting of environmental envelopes, logistic regression and categorical rules. This approach was intended to capture complex heterogeneous types of relationships of species to the environment, together with robustly handling the different types of environmental data [SP99].

Although most machine learning methods applied to niche modeling result in estimates of a probability distribution, they are problematic to interpret in familiar terms, as the form of the model is not a simple envelope of linear model. Another drawback was that some required multiple runs and are computationally intensive – a potentially serious limitation if in addition they do not scale well to large numbers of variables. Nevertheless, the development of these machine-learning methods has progressed and many are giving very good results exceeding the classical approaches [EGA+06].

4.7   Data mining

Data mining is the automated search for patterns in large amounts of data. A couple of aspects of niche modeling make data mining potentially useful. Firstly, as often little is known about the factors determining species' distributions, we don't know what factors will be most accurate at predicting the species. Because of this uncertainty, we can't always use the same variables, such as annual averages of temperature and rainfall, and expect to get a good model.

For example species in freshwater and marine environments are not well modeled by annual climate factors, and as the popularity of niche modeling grows more entities in exotic environments will be of interest. Data mining algorithms are designed to test a large number of datasets as potential candidates for models. Secondly there is a lot more data available now than there was - the subject of the following chapter.

The goal of a data mining approach to niche modeling is for minimal assumptions to be made about the type of variables and the form of the probability distribution that can potentially be used in a model. An approach allowing virtually any variable to be used, necessitates the generalization of the notion of environmental space to include countably infinite environmental variables. That there are potentially infinitely variables is clear, even though at any time only finitely many have been recorded. So the niche space X becomes:

$$\prod_{i \in I} X_i \text{ or } X_1 \times ... \times X_n \times ...$$

where I is the set of integers. To a large extent the only difference between standard analysis and data mining is the number of independent variables. Defining niche modeling as an infinite product highlights this in a theoretical way with practical implications. For example, models developed by fitting all variables simultaneously cannot really be viable in a space of infinite variables. In practice, the datasets generated by such a procedure would be too large for computer memory systems.

Data mining is often distinguished from conventional niche modeling in that a sequential approach to including variables in the model is used. It may also be the said that data mining generally uses non-parametric methods to robustly discover information within a large number of variables with a range of types of distributions.

4.7.1 Decision trees

One of the most popular approaches used in data mining is the induction of decision trees, based on the sequential partitioning of datasets on individual variables. Mentioned before under machine learning methods, decision tree methods have continued to be improved with the use of more complex algorithms for improving robustness. For example, one of the more important is a recent classification method from machine learning that uses a process called *boosting*, a way of combining the performance of many 'weak' classifiers to produce a powerful 'committee' [FHT98].

4.7.2 Clustering

Another approach to data mining is clustering, which has broad appeal as an exploratory method of data analysis in many fields [JMF99]. Methods such as k-means quantize variables into a discrete number of groups, and characterize the points in the groups by representative features, such as the group centroids. In comparison to more heuristic methods, the statistics of k-means and decision tree methods are well understood.

WhyWhere data-mining approach to niche modeling [Sto06] uses clustering. Here an image processing method derives the categories from up to three environmental variables, characterized as the list of reduced colors. Efficient approximate implementations of k-means are used for the color reduction based on Heckbert's median cut. Used in GIF and other image formats to compress their size, Heckbert's algorithm has been proven to give efficient, though not necessarily optimal results for images [Hec82].

In clustering approaches, probabilities for prediction at a specific point are derived from a single probability at each cluster. These can simply be the cluster the point belongs to, or a weighted sum of probabilities at a number of clusters. In WhyWhere the probabilities of presence or absence are calculated from the proportion of occurrences of the points in a group relative to the proportion of environmental values in that category.

4.7.3 Comparison

What distinguishes features of decision trees and k-means from logistic regression, environmental envelope and other more conventional methods?

The first distinguishing feature is the capacity for representation in parts, rather than as a whole, connected space. Secondly, data mining has the capacity to examine, if not simultaneously, large numbers of variables. These capacities address the reality of data analysis in their real world, stressing optimized performance. In contrast, conventional methods tend to stress methods that in some way express ones' understanding of the theoretical structure of the domain.

One could say that data mining tends towards pragmatism, and conventional methods tend towards idealism.

However, features such as lack of connectedness can have ecological meaning. For example, species may be found in more than one type of vegetation, because it is a widely distributed generalist, relies on different vegetation types for different resources (e.g. food and nesting) or simply because of the particulars of the classification scheme. This species would have a niche model using vegetation variables that are separable and not connected, i.e. there is a broken path between two vegetation types.

An example using continuous variables is a species composed of two distinct populations – a very common situation with some physiologically plastic genera, e.g. Notophagus, and many widespread bird species. Finally, there is the precedent of community ecology, which makes extensive use of clustering techniques for defining communities, based in notions of separability. So clearly there is an ecological motivation to admit separable models.

4.8   Post-Hutchinsonian niche

The possibility of defining niches in the light of these developments is equally interesting. In the Hutchinsonian definition a niche is defined on all 'ecologically relevant variables' however defined. The simple construction of the niche is an environmental envelope N containing all the points of occurrence of the species in B.

What happens when the number of environmental variables is potentially infinite? The definition 'ecologically relevant' does not specify how to exclude variables from the environmental envelope, an infinite dimensional hypervolume results. This is problematic as it is not constructible. Constructing a Hutchinsonian niche would require specifying conditions on an infinite number of datasets. While the Axiom of Choice states this is possible, which suggests an arbitrary Cartesian product of non-empty sets is itself non-empty, it is not algorithmically possible.

If the niche is defined slightly differently, as a mapping from an infinite number of variables to a finite number, constructability of the niche is retained.

In theoretical terms, a Hutchinsonian niche over infinite variables based on a box topology is a box of infinite dimension. However, an alternative approach to defining a niche j would be to use a projection map:

$$\pi_j : \prod_{i \in I} X_i \longrightarrow X_j$$

In this case a point in the niche is represented by an n-tuple $(x_1, ..., x_n)$ representing a family of elements of X the infinite space.

A topological space defined in this way is known as a product space or product topology. While the product and the box topology are very similar, and identical over finite variables, the product topology has more desirable properties over infinite variables that make it more widely used in modern topology [Mun75].

The limitation of niche definition to finite dimensions is also consistent with the usual strategies for reducing overfitting, such as stepwise addition or deletion of variables in a logistic regression or ℓ_1-regularization in Maxent, which only include in models the most important features [PAS06]. These strategies are typically justified statistically, e.g. by divergence of the finite sample of data from the true distribution being sought. Here, we show topological properties such as constructability, and continuity are preserved by defining the niche with finite dimensionality.

4.8.1 Product space

Formalizing the niche as a product space may be a worthwhile upgrade to Hutchinsonian formalism of the niche.

A small proof of continuity of the product space illustrates the value of a product topology over the box topology. One difference between the product and the box topology is that in the product topology, the continuity of a function in the Cartesian product is guaranteed by the continuity of functions on each of the components. Continuity of a function from physical space B to an abstract n-dimensional space N is the relationship such that for any small range change ΔB there may be a corresponding small change in the niche of the species ΔN, and vice versa.

More formally, a function $f : B \longrightarrow N$ is continuous if for each subset X of N, the set defined by the inverse function $f^{-1}(Y)$ is an open set in B.

Here is an example of a niche construct that is not continuous in a box topology but continuous in a product topology. Consider the approach commonly used to construct an environmental envelope where the niche formed by the functions $f_i(r)$ defines a interval X_i of the environmental variable enclosing a proportion $r < 1$ of the points of occurrence of species in the geographic space B. For example, the environmental interval may contain 95% of the cells b where the species occurs. The functions f_i are continuous because each interval X_i defined by $f_i(r)$ is a non-empty set containing r points of occurrence in B.

The n dimensional coordinate function is simply defined on each of the variables:

$$f(r) = (f_1(r) \times \ldots) = (X_1, \ldots)$$

The n-dimensional hypervolume, or volume in the box topology, will be given by:

$$V = r^n.$$

However, the volume V will converge to zero in the infinite limit:

$$\lim_{n \to \infty} r^n = 0$$

In the inverse function $f^{-1} : N \longrightarrow B$ the box with zero volume is the empty set, containing no occurrences of the species, or zero geographical range. Being empty, the set is not open in B. So while each of the component functions f_i is continuous, with each $f_i^{-1}(X_i)$ defining r occurrences of b, the function f under the box topology is not continuous. The volume of the niche is zero in the infinite limit, meaning the projected range in B from N is empty despite being non-empty in each of the component functions.

This example shows theoretically the origin of the progressive reduction in ecological area with inclusion of each variable as evidenced experimentally in envelope models by Beaumont [BHP05]. The function f is continuous in the product topology, however, because the number of component functions are finite, the limit under infinite environmental variables is not zero, and the geographic range of the species does not vanish.

This example illustrates the advantage of defining the niche on a finite projection of potentially infinite variables. In contrast, the Hutchinsonian niche definition defining the niche as a hypervolume on all ecologically relevant variables, which can be potentially infinite, leads to undesirable topological properties.

4.9 Summary

Set theory helps to identify the basic assumptions underlying niche modeling, and show some relationships between these assumptions, methodology and representation of intuitive understanding of the concept of a niche. The chapter concludes with a proof of the lack of continuity of the standard definition of environmental envelopes over the box topology, and argues for definition of the niche in the product topology.

Chapter 5

Environmental data collections

Previous chapters highlighted the analysis of data, in particular the possibility of a potentially infinite number of environmental variables. Here we consider some of the large number of datasets that might be used to develop niche models, and the use of archives to manage them.

In the past, a small number of specially prepared, primarily climatic variables, such as annual average temperature and rainfall, comprised the variables in a niche model. This approach is sub-optimal for a number of reasons:

- ignores a large number of potentially relevant variables,

- potential for errors in individual preparation of data,

- duplication of effort, and

- data are vulnerable to corruption and loss.

Niche models in the future could be prepared much more efficiently from large numbers of data sets acquired from well maintained databases called data archives. The aim of a data archive is to consistently provide these environmental correlates from a massive archive of data, through efficient interaction with modeling systems. Niche models developed using such archives have many advantages, including:

- access to greater numbers of variables,

- consistent verifiable quality,

- studies replicable by others,

- faster, cheaper model development, and

- safe backup.

5.1   Datasets

Niche modelers typically restrict models to few variables, mainly related to temperature and rainfall. Although these variables do appear important to most species, this approach ignores many potential variables that could be relevant, if not better predictors of the distribution of the species.

The following lists some of the major resources for ecological niche modeling. The main purpose of listing these files is not to be comprehensive, but to illustrate how many variables could regarded as potentially 'ecologically relevant' to a species.

Examples of possible correlates include monthly temperatures and rainfall, functions of these variables such as standard deviations, and those related to water availability and evapotransporation, soil and vegetation habitat conditions, and topography. Extended into the marine environment, each variable potentially exists in three dimensions. Combine this with remote sensing data, the existence of alternative versions of variables, different scales, and temporal factors such as time and duration and the number of variables that a researcher might want to examine for potential correlates expands rapidly.

Below is a listing of over one thousand variables currently in the WhyWhere archive. The entire list is printed so the reader can get an impression of the size of one thousand files, many of which on examination could be regarded as ecologically relevant to an arbitrary marine or terrestrial species. Particularly as this is only a subset of data that does not include many satellite datasets, it supports the claim that the number of ecologically relevant datasets is very large, and could be regarded as potentially infinite.

1. a00an1.1.txt World Ocean Atlas 2001 1 degree annual analyzed mean - apparent oxygen utilization at depth 0 metre

2. a00an1.10.txt World Ocean Atlas 2001 1 degree annual analyzed mean - apparent oxygen utilization at depth 200 metre

3. a00an1.11.txt World Ocean Atlas 2001 1 degree annual analyzed mean - apparent oxygen utilization at depth 250 metre

4. a00an1.12.txt World Ocean Atlas 2001 1 degree annual analyzed mean - apparent oxygen utilization at depth 300 metre

5. a00an1.13.txt World Ocean Atlas 2001 1 degree annual analyzed mean - apparent oxygen utilization at depth 400 metre

6. a00an1.14.txt World Ocean Atlas 2001 1 degree annual analyzed mean - apparent oxygen utilization at depth 500 metre

7. a00an1.15.txt World Ocean Atlas 2001 1 degree annual analyzed mean - apparent oxygen utilization at depth 600 metre

8. a00an1.16.txt World Ocean Atlas 2001 1 degree annual analyzed mean - apparent oxygen utilization at depth 700 metre

9. a00an1.17.txt World Ocean Atlas 2001 1 degree annual analyzed mean - apparent oxygen utilization at depth 800 metre

10. a00an1.18.txt World Ocean Atlas 2001 1 degree annual analyzed mean - apparent oxygen utilization at depth 900 metre

11. a00an1.19.txt World Ocean Atlas 2001 1 degree annual analyzed mean - apparent oxygen utilization at depth 1000 metre

12. a00an1.2.txt World Ocean Atlas 2001 1 degree annual analyzed mean - apparent oxygen utilization at depth 10 metre

13. a00an1.20.txt World Ocean Atlas 2001 1 degree annual analyzed mean - apparent oxygen utilization at depth 1100 metre

14. a00an1.21.txt World Ocean Atlas 2001 1 degree annual analyzed mean - apparent oxygen utilization at depth 1200 metre

15. a00an1.22.txt World Ocean Atlas 2001 1 degree annual analyzed mean - apparent oxygen utilization at depth 1300 metre

16. a00an1.23.txt World Ocean Atlas 2001 1 degree annual analyzed mean - apparent oxygen utilization at depth 1400 metre

17. a00an1.24.txt World Ocean Atlas 2001 1 degree annual analyzed mean - apparent oxygen utilization at depth 1500 metre

18. a00an1.25.txt World Ocean Atlas 2001 1 degree annual analyzed mean - apparent oxygen utilization at depth 1750 metre

19. a00an1.26.txt World Ocean Atlas 2001 1 degree annual analyzed mean - apparent oxygen utilization at depth 2000 metre

20. a00an1.27.txt World Ocean Atlas 2001 1 degree annual analyzed mean - apparent oxygen utilization at depth 2500 metre

21. a00an1.28.txt World Ocean Atlas 2001 1 degree annual analyzed mean - apparent oxygen utilization at depth 3000 metre

22. a00an1.29.txt World Ocean Atlas 2001 1 degree annual analyzed mean - apparent oxygen utilization at depth 3500 metre

23. a00an1.3.txt World Ocean Atlas 2001 1 degree annual analyzed mean - apparent oxygen utilization at depth 20 metre

24. a00an1.30.txt World Ocean Atlas 2001 1 degree annual analyzed mean - apparent oxygen utilization at depth 4000 metre

25. a00an1.31.txt World Ocean Atlas 2001 1 degree annual analyzed mean - apparent oxygen utilization at depth 4500 metre

26. a00an1.32.txt World Ocean Atlas 2001 1 degree annual analyzed mean - apparent oxygen utilization at depth 5000 metre

27. a00an1.33.txt World Ocean Atlas 2001 1 degree annual analyzed mean - apparent oxygen utilization at depth 5500 metre

28. a00an1.4.txt World Ocean Atlas 2001 1 degree annual analyzed mean - apparent oxygen utilization at depth 30 metre

29. a00an1.5.txt World Ocean Atlas 2001 1 degree annual analyzed mean - apparent oxygen utilization at depth 50 metre

30. a00an1.6.txt World Ocean Atlas 2001 1 degree annual analyzed mean - apparent oxygen utilization at depth 75 metre

31. a00an1.7.txt World Ocean Atlas 2001 1 degree annual analyzed mean - apparent oxygen utilization at depth 100 metre

32. a00an1.8.txt World Ocean Atlas 2001 1 degree annual analyzed mean - apparent oxygen utilization at depth 125 metre

33. a00an1.9.txt World Ocean Atlas 2001 1 degree annual analyzed mean - apparent oxygen utilization at depth 150 metre

34. a00sd1.1.txt World Ocean Atlas 2001 1 degree standard deviation - apparent oxygen utilization at depth 0 metre

35. a00sd1.10.txt World Ocean Atlas 2001 1 degree standard deviation - apparent oxygen utilization at depth 200 metre

36. a00sd1.11.txt World Ocean Atlas 2001 1 degree standard deviation - apparent oxygen utilization at depth 250 metre

37. a00sd1.12.txt World Ocean Atlas 2001 1 degree standard deviation - apparent oxygen utilization at depth 300 metre

38. a00sd1.13.txt World Ocean Atlas 2001 1 degree standard deviation - apparent oxygen utilization at depth 400 metre

39. a00sd1.14.txt World Ocean Atlas 2001 1 degree standard deviation - apparent oxygen utilization at depth 500 metre

40. a00sd1.15.txt World Ocean Atlas 2001 1 degree standard deviation - apparent oxygen utilization at depth 600 metre

41. a00sd1.16.txt World Ocean Atlas 2001 1 degree standard deviation - apparent oxygen utilization at depth 700 metre

42. a00sd1.17.txt World Ocean Atlas 2001 1 degree standard deviation - apparent oxygen utilization at depth 800 metre

43. a00sd1.18.txt World Ocean Atlas 2001 1 degree standard deviation - apparent oxygen utilization at depth 900 metre

44. a00sd1.19.txt World Ocean Atlas 2001 1 degree standard deviation - apparent oxygen utilization at depth 1000 metre

45. a00sd1.2.txt World Ocean Atlas 2001 1 degree standard deviation - apparent oxygen utilization at depth 10 metre

46. a00sd1.20.txt World Ocean Atlas 2001 1 degree standard deviation - apparent oxygen utilization at depth 1100 metre

47. a00sd1.21.txt World Ocean Atlas 2001 1 degree standard deviation - apparent oxygen utilization at depth 1200 metre

48. a00sd1.22.txt World Ocean Atlas 2001 1 degree standard deviation - apparent oxygen utilization at depth 1300 metre

49. a00sd1.23.txt World Ocean Atlas 2001 1 degree standard deviation - apparent oxygen utilization at depth 1400 metre

50. a00sd1.24.txt World Ocean Atlas 2001 1 degree standard deviation - apparent oxygen utilization at depth 1500 metre

51. a00sd1.25.txt World Ocean Atlas 2001 1 degree standard deviation - apparent oxygen utilization at depth 1750 metre

52. a00sd1.26.txt World Ocean Atlas 2001 1 degree standard deviation - apparent oxygen utilization at depth 2000 metre

53. a00sd1.27.txt World Ocean Atlas 2001 1 degree standard deviation - apparent oxygen utilization at depth 2500 metre

54. a00sd1.28.txt World Ocean Atlas 2001 1 degree standard deviation - apparent oxygen utilization at depth 3000 metre

55. a00sd1.29.txt World Ocean Atlas 2001 1 degree standard deviation - apparent oxygen utilization at depth 3500 metre

56. a00sd1.3.txt World Ocean Atlas 2001 1 degree standard deviation - apparent oxygen utilization at depth 20 metre

57. a00sd1.30.txt World Ocean Atlas 2001 1 degree standard deviation - apparent oxygen utilization at depth 4000 metre

58. a00sd1.31.txt World Ocean Atlas 2001 1 degree standard deviation - apparent oxygen utilization at depth 4500 metre

59. a00sd1.32.txt World Ocean Atlas 2001 1 degree standard deviation - apparent oxygen utilization at depth 5000 metre

60. a00sd1.33.txt World Ocean Atlas 2001 1 degree standard deviation - apparent oxygen utilization at depth 5500 metre

61. a00sd1.4.txt World Ocean Atlas 2001 1 degree standard deviation - apparent oxygen utilization at depth 30 metre

62. a00sd1.5.txt World Ocean Atlas 2001 1 degree standard deviation - apparent oxygen utilization at depth 50 metre

63. a00sd1.6.txt World Ocean Atlas 2001 1 degree standard deviation - apparent oxygen utilization at depth 75 metre

64. a00sd1.7.txt World Ocean Atlas 2001 1 degree standard deviation - apparent oxygen utilization at depth 100 metre

65. a00sd1.8.txt World Ocean Atlas 2001 1 degree standard deviation - apparent oxygen utilization at depth 125 metre

66. a00sd1.9.txt World Ocean Atlas 2001 1 degree standard deviation - apparent oxygen utilization at depth 150 metre

67. alt.txt Altitude

68. alt_2.5m.txt alt 2.5 minutes

69. bio_1.txt Annual Mean Temperature

70. bio_10.txt Mean Temperature of Warmest Quarter

71. bio_10_2.5m.txt bio_10 2.5 minutes

72. bio_11.txt Mean Temperature of Coldest Quarter

73. bio_11_2.5m.txt bio_11 2.5 minutes

74. bio_12.txt Annual Precipitation

75. bio_12_2.5m.txt bio_12 2.5 minutes

76. bio_13.txt Precipitation of Wettest Month

77. bio_13_2.5m.txt bio_13 2.5 minutes

78. bio_14.txt Precipitation of Driest Month

79. bio_14_2.5m.txt bio_14 2.5 minutes

80. bio_15.txt Precipitation Seasonality (Coefficient of Variation)

81. bio_15_2.5m.txt bio_15 2.5 minutes

82. bio_16.txt Precipitation of Wettest Quarter

83. bio_16_2.5m.txt bio_16 2.5 minutes

84. bio_17.txt Precipitation of Driest Quarter

85. bio_17_2.5m.txt bio_17 2.5 minutes

86. bio_18.txt Precipitation of Warmest Quarter

87. bio_18_2.5m.txt bio_18 2.5 minutes

88. bio_19.txt Precipitation of Coldest Quarter

89. bio_19_2.5m.txt bio_19 2.5 minutes

90. bio_1_2.5m.txt bio_1 2.5 minutes

91. bio_2.txt Mean Diurnal Range (Mean of monthly (max temp - min temp))

92. bio_2_2.5m.txt bio_2 2.5 minutes

93. bio_3.txt Isothermality (P2/P7) (* 100)

94. bio_3_2.5m.txt bio_3 2.5 minutes

95. bio_4.txt Temperature Seasonality (standard deviation *100)

96. bio_4_2.5m.txt bio_4 2.5 minutes

97. bio_5.txt Max Temperature of Warmest Month

98. bio_5_2.5m.txt bio_5 2.5 minutes

99. bio_6.txt Min Temperature of Coldest Month

100. bio_6_2.5m.txt bio_6 2.5 minutes

101. bio_7.txt Temperature Annual Range (P5-P6)

102. bio_7_2.5m.txt bio_7 2.5 minutes

103. bio_8.txt Mean Temperature of Wettest Quarter

104. bio_8_2.5m.txt bio_8 2.5 minutes

105. bio_9.txt Mean Temperature of Driest Quarter

106. bio_9_2.5m.txt bio_9 2.5 minutes

107. broadleaf.txt Continuoous field data - broadleaf

108. c00an1.1.txt World Ocean Atlas 2001 1 degree annual analyzed mean - chlorophyll at depth 0 metre

109. c00an1.2.txt World Ocean Atlas 2001 1 degree annual analyzed mean - chlorophyll at depth 10 metre

110. c00an1.3.txt World Ocean Atlas 2001 1 degree annual analyzed mean - chlorophyll at depth 20 metre

111. c00an1.4.txt World Ocean Atlas 2001 1 degree annual analyzed mean - chlorophyll at depth 30 metre

112. c00an1.5.txt World Ocean Atlas 2001 1 degree annual analyzed mean - chlorophyll at depth 50 metre

113. c00an1.6.txt World Ocean Atlas 2001 1 degree annual analyzed mean - chlorophyll at depth 75 metre

114. c00an1.7.txt World Ocean Atlas 2001 1 degree annual analyzed mean - chlorophyll at depth 100 metre

115. c00sd1.1.txt World Ocean Atlas 2001 1 degree standard deviation - chlorophyll at depth 0 metre

116. c00sd1.2.txt World Ocean Atlas 2001 1 degree standard deviation - chlorophyll at depth 10 metre

117. c00sd1.3.txt World Ocean Atlas 2001 1 degree standard deviation - chlorophyll at depth 20 metre

118. c00sd1.4.txt World Ocean Atlas 2001 1 degree standard deviation - chlorophyll at depth 30 metre

119. c00sd1.5.txt World Ocean Atlas 2001 1 degree standard deviation - chlorophyll at depth 50 metre

120. c00sd1.6.txt World Ocean Atlas 2001 1 degree standard deviation - chlorophyll at depth 75 metre

121. c00sd1.7.txt World Ocean Atlas 2001 1 degree standard deviation - chlorophyll at depth 100 metre

122. CVS 122. deciduous.txt Continuous field data - deciduous

123. etopo2-marine.txt etopo2

124. etopo2-terr.txt etopo2

125. etopo2.txt etopo2

126. evergreen.txt Continuous field data - evergreen

127. fnocazm.txt Navy Terrain Data–Direction of Ridges (degrees X 10)

128. fnocmax.txt Navy Terrain Data–Maximum Elevation (meters)

129. fnocmin.txt Navy Terrain Data–Minimum Elevation (meters)

130. fnocmod.txt Navy Terrain Data–Modal Elevation (meters)

131. fnococm.txt Ocean Mask (produced from Navy Terrain data)

132. fnocpt.txt Navy Terrain Data–Primary Surface Type Codes

133. fnocrdg.txt Navy Terrain Data–Number of Significant Ridges

134. fnocst.txt Navy Terrain Data–Secondary Surface Type Codes

135. fnocurb.txt Navy Terrain Data–Percent Urban Cover

136. fnocwat.txt Navy Terrain Data–Percent Water Cover

137. hydro_aspect.txt Aspect Data

138. hydro_dem.txt Elevation Data

139. hydro_flowacc.txt Flow Accumulation

70

Niche Modeling

140. hydro_flowdir.txt Flow direction
141. hydro_slope.txt Slope Data
142. hydro_topoind.txt Compound Topographic Index Data
143. i00an1.1.txt World Ocean Atlas 2001 1 degree annual analyzed mean - silicate at depth 0 metre
144. i00an1.10.txt World Ocean Atlas 2001 1 degree annual analyzed mean - silicate at depth 200 metre
145. i00an1.11.txt World Ocean Atlas 2001 1 degree annual analyzed mean - silicate at depth 250 metre
146. i00an1.12.txt World Ocean Atlas 2001 1 degree annual analyzed mean - silicate at depth 300 metre
147. i00an1.13.txt World Ocean Atlas 2001 1 degree annual analyzed mean - silicate at depth 400 metre
148. i00an1.14.txt World Ocean Atlas 2001 1 degree annual analyzed mean - silicate at depth 500 metre
149. i00an1.15.txt World Ocean Atlas 2001 1 degree annual analyzed mean - silicate at depth 600 metre
150. i00an1.16.txt World Ocean Atlas 2001 1 degree annual analyzed mean - silicate at depth 700 metre
151. i00an1.17.txt World Ocean Atlas 2001 1 degree annual analyzed mean - silicate at depth 800 metre
152. i00an1.18.txt World Ocean Atlas 2001 1 degree annual analyzed mean - silicate at depth 900 metre
153. i00an1.19.txt World Ocean Atlas 2001 1 degree annual analyzed mean - silicate at depth 1000 metre
154. i00an1.2.txt World Ocean Atlas 2001 1 degree annual analyzed mean - silicate at depth 10 metre
155. i00an1.20.txt World Ocean Atlas 2001 1 degree annual analyzed mean - silicate at depth 1100 metre
156. i00an1.21.txt World Ocean Atlas 2001 1 degree annual analyzed mean - silicate at depth 1200 metre
157. i00an1.22.txt World Ocean Atlas 2001 1 degree annual analyzed mean - silicate at depth 1300 metre
158. i00an1.23.txt World Ocean Atlas 2001 1 degree annual analyzed mean - silicate at depth 1400 metre
159. i00an1.24.txt World Ocean Atlas 2001 1 degree annual analyzed mean - silicate at depth 1500 metre
160. i00an1.25.txt World Ocean Atlas 2001 1 degree annual analyzed mean - silicate at depth 1750 metre
161. i00an1.26.txt World Ocean Atlas 2001 1 degree annual analyzed mean - silicate at depth 2000 metre
162. i00an1.27.txt World Ocean Atlas 2001 1 degree annual analyzed mean - silicate at depth 2500 metre
163. i00an1.28.txt World Ocean Atlas 2001 1 degree annual analyzed mean - silicate at depth 3000 metre
164. i00an1.29.txt World Ocean Atlas 2001 1 degree annual analyzed mean - silicate at depth 3500 metre
165. i00an1.3.txt World Ocean Atlas 2001 1 degree annual analyzed mean - silicate at depth 20 metre
166. i00an1.30.txt World Ocean Atlas 2001 1 degree annual analyzed mean - silicate at depth 4000 metre
167. i00an1.31.txt World Ocean Atlas 2001 1 degree annual analyzed mean - silicate at depth 4500 metre
168. i00an1.32.txt World Ocean Atlas 2001 1 degree annual analyzed mean - silicate at depth 5000 metre
169. i00an1.33.txt World Ocean Atlas 2001 1 degree annual analyzed mean - silicate at depth 5500 metre
170. i00an1.4.txt World Ocean Atlas 2001 1 degree annual analyzed mean - silicate at depth 30 metre
171. i00an1.5.txt World Ocean Atlas 2001 1 degree annual analyzed mean - silicate at depth 50 metre
172. i00an1.6.txt World Ocean Atlas 2001 1 degree annual analyzed mean - silicate at depth 75 metre
173. i00an1.7.txt World Ocean Atlas 2001 1 degree annual analyzed mean - silicate at depth 100 metre
174. i00an1.8.txt World Ocean Atlas 2001 1 degree annual analyzed mean - silicate at depth 125 metre
175. i00an1.9.txt World Ocean Atlas 2001 1 degree annual analyzed mean - silicate at depth 150 metre
176. i00sd1.1.txt World Ocean Atlas 2001 1 degree standard deviation - silicate at depth 0 metre
177. i00sd1.10.txt World Ocean Atlas 2001 1 degree standard deviation - silicate at depth 200 metre
178. i00sd1.11.txt World Ocean Atlas 2001 1 degree standard deviation - silicate at depth 250 metre
179. i00sd1.12.txt World Ocean Atlas 2001 1 degree standard deviation - silicate at depth 300 metre
180. i00sd1.13.txt World Ocean Atlas 2001 1 degree standard deviation - silicate at depth 400 metre
181. i00sd1.14.txt World Ocean Atlas 2001 1 degree standard deviation - silicate at depth 500 metre
182. i00sd1.15.txt World Ocean Atlas 2001 1 degree standard deviation - silicate at depth 600 metre
183. i00sd1.16.txt World Ocean Atlas 2001 1 degree standard deviation - silicate at depth 700 metre
184. i00sd1.17.txt World Ocean Atlas 2001 1 degree standard deviation - silicate at depth 800 metre
185. i00sd1.18.txt World Ocean Atlas 2001 1 degree standard deviation - silicate at depth 900 metre
186. i00sd1.19.txt World Ocean Atlas 2001 1 degree standard deviation - silicate at depth 1000 metre
187. i00sd1.2.txt World Ocean Atlas 2001 1 degree standard deviation - silicate at depth 10 metre
188. i00sd1.20.txt World Ocean Atlas 2001 1 degree standard deviation - silicate at depth 1100 metre
189. i00sd1.21.txt World Ocean Atlas 2001 1 degree standard deviation - silicate at depth 1200 metre
190. i00sd1.22.txt World Ocean Atlas 2001 1 degree standard deviation - silicate at depth 1300 metre
191. i00sd1.23.txt World Ocean Atlas 2001 1 degree standard deviation - silicate at depth 1400 metre
192. i00sd1.24.txt World Ocean Atlas 2001 1 degree standard deviation - silicate at depth 1500 metre
193. i00sd1.25.txt World Ocean Atlas 2001 1 degree standard deviation - silicate at depth 1750 metre

194. i00sd1.26.txt World Ocean Atlas 2001 1 degree standard deviation - silicate at depth 2000 metre
195. i00sd1.27.txt World Ocean Atlas 2001 1 degree standard deviation - silicate at depth 2500 metre
196. i00sd1.28.txt World Ocean Atlas 2001 1 degree standard deviation - silicate at depth 3000 metre
197. i00sd1.29.txt World Ocean Atlas 2001 1 degree standard deviation - silicate at depth 3500 metre
198. i00sd1.3.txt World Ocean Atlas 2001 1 degree standard deviation - silicate at depth 20 metre
199. i00sd1.30.txt World Ocean Atlas 2001 1 degree standard deviation - silicate at depth 4000 metre
200. i00sd1.31.txt World Ocean Atlas 2001 1 degree standard deviation - silicate at depth 4500 metre
201. i00sd1.32.txt World Ocean Atlas 2001 1 degree standard deviation - silicate at depth 5000 metre
202. i00sd1.33.txt World Ocean Atlas 2001 1 degree standard deviation - silicate at depth 5500 metre
203. i00sd1.4.txt World Ocean Atlas 2001 1 degree standard deviation - silicate at depth 30 metre
204. i00sd1.5.txt World Ocean Atlas 2001 1 degree standard deviation - silicate at depth 50 metre
205. i00sd1.6.txt World Ocean Atlas 2001 1 degree standard deviation - silicate at depth 75 metre
206. i00sd1.7.txt World Ocean Atlas 2001 1 degree standard deviation - silicate at depth 100 metre
207. i00sd1.8.txt World Ocean Atlas 2001 1 degree standard deviation - silicate at depth 125 metre
208. i00sd1.9.txt World Ocean Atlas 2001 1 degree standard deviation - silicate at depth 150 metre
209. Laws_Export_Period_9709_9808.txt Laws Export in Period 9709 9808
210. lccld01.txt Leemans and Cramer January Cloudiness (% Sunshine)
211. lccld02.txt Leemans and Cramer February Cloudiness (% Sunshine)
212. lccld03.txt Leemans and Cramer March Cloudiness (% Sunshine)
213. lccld04.txt Leemans and Cramer April Cloudiness (% Sunshine)
214. lccld05.txt Leemans and Cramer May Cloudiness (% Sunshine)
215. lccld06.txt Leemans and Cramer June Cloudiness (% Sunshine)
216. lccld07.txt Leemans and Cramer July Cloudiness (% Sunshine)
217. lccld08.txt Leemans and Cramer August Cloudiness (% Sunshine)
218. lccld09.txt Leemans and Cramer September Cloudiness (% Sunshine)
219. lccld10.txt Leemans and Cramer October Cloudiness (% Sunshine)
220. lccld11.txt Leemans and Cramer November Cloudiness (% Sunshine)
221. lccld12.txt Leemans and Cramer December Cloudiness (% Sunshine)
222. lcprc01.txt Leemans and Cramer January Precipitation (mm/month)
223. lcprc02.txt Leemans and Cramer February Precipitation (mm/month)
224. lcprc03.txt Leemans and Cramer March Precipitation (mm/month)
225. lcprc04.txt Leemans and Cramer April Precipitation (mm/month)
226. lcprc05.txt Leemans and Cramer May Precipitation (mm/month)
227. lcprc06.txt Leemans and Cramer June Precipitation (mm/month)
228. lcprc07.txt Leemans and Cramer July Precipitation (mm/month)
229. lcprc08.txt Leemans and Cramer August Precipitation (mm/month)
230. lcprc09.txt Leemans and Cramer September Precipitation (mm/month)
231. lcprc10.txt Leemans and Cramer October Precipitation (mm/month)
232. lcprc11.txt Leemans and Cramer November Precipitation (mm/month)
233. lcprc12.txt Leemans and Cramer December Precipitation (mm/month)
234. lctmp01.txt Leemans and Cramer January Temperature (0.1C)
235. lctmp02.txt Leemans and Cramer February Temperature (0.1C)
236. lctmp03.txt Leemans and Cramer March Temperature (0.1C)
237. lctmp04.txt Leemans and Cramer April Temperature (0.1C)
238. lctmp05.txt Leemans and Cramer May Temperature (0.1C)
239. lctmp06.txt Leemans and Cramer June Temperature (0.1C)
240. lctmp07.txt Leemans and Cramer July Temperature (0.1C)
241. lctmp08.txt Leemans and Cramer August Temperature (0.1C)
242. lctmp09.txt Leemans and Cramer September Temperature (0.1C)
243. lctmp10.txt Leemans and Cramer October Temperature (0.1C)
244. lctmp11.txt Leemans and Cramer November Temperature (0.1C)
245. lctmp12.txt Leemans and Cramer December Temperature (0.1C)
246. lhold.txt Leemans' Holdridge Life Zones Classification
247. lholdag.txt Leemans' Holdridge Life Zones Aggregated Classification

248. lmfcaml.txt Lerner et al Camel Density (1/Km^2)
249. lmfcarb.txt Lerner et al Caribou Density (1/Km^2)
250. lmfcow.txt Lerner et al Dairy and Non-Dairy Cattle Density (1/Km^2)
251. lmfdcow.txt Lerner et al Dairy Cow Density (1/Km^2)
252. lmfgoat.txt Lerner et al Goat Density (1/Km^2)
253. lmfhors.txt Lerner et al Horse Density (1/Km^2)
254. lmfmeth.txt Lerner et al Annual Methane Emission (Kg/Km^2)
255. lmfncow.txt Lerner et al Non-Dairy Cattle Density (1/Km^2)
256. lmfpig.txt Lerner et al Pig Density (1/Km^2)
257. lmfshep.txt Lerner et al Sheep Density (1/Km^2)
258. lmfwbuf.txt Lerner et al Water Buffalo Density (1/Km^2)
259. lwcpr00.txt Legates & Willmott Annual Corrected Precipitation (mm/year)
260. lwcpr01.txt Legates & Willmott January Corrected Precipitation (mm/month)
261. lwcpr02.txt Legates & Willmott February Corrected Precipitation (mm/month)
262. lwcpr03.txt Legates & Willmott March Corrected Precipitation (mm/month)
263. lwcpr04.txt Legates & Willmott April Corrected Precipitation (mm/month)
264. lwcpr05.txt Legates & Willmott May Corrected Precipitation (mm/month)
265. lwcpr06.txt Legates & Willmott June Corrected Precipitation (mm/month)
266. lwcpr07.txt Legates & Willmott July Corrected Precipitation (mm/month)
267. lwcpr08.txt Legates & Willmott August Corrected Precipitation (mm/month)
268. lwcpr09.txt Legates & Willmott September Corrected Precipitation (mm/month)
269. lwcpr10.txt Legates & Willmott October Corrected Precipitation (mm/month)
270. lwcpr11.txt Legates & Willmott November Corrected Precipitation (mm/month)
271. lwcpr12.txt Legates & Willmott December Corrected Precipitation (mm/month)
272. lwcsd00.txt Legates & Willmott Annual Corrected Precipitation (std. dev.)
273. lwcsd01.txt Legates & Willmott January Corrected Precipitation (std. dev.)
274. lwcsd02.txt Legates & Willmott February Corrected Precipitation (std. dev.)
275. lwcsd03.txt Legates & Willmott March Corrected Precipitation (std. dev.)
276. lwcsd04.txt Legates & Willmott April Corrected Precipitation (std. dev.)
277. lwcsd05.txt Legates & Willmott May Corrected Precipitation (std. dev.)
278. lwcsd06.txt Legates & Willmott June Corrected Precipitation (std. dev.)
279. lwcsd07.txt Legates & Willmott July Corrected Precipitation (std. dev.)
280. lwcsd08.txt Legates & Willmott August Corrected Precipitation (std. dev.)
281. lwcsd09.txt Legates & Willmott September Corrected Precipitation (std. dev.)
282. lwcsd10.txt Legates & Willmott October Corrected Precipitation (std. dev.)
283. lwcsd11.txt Legates & Willmott November Corrected Precipitation (std. dev.)
284. lwcsd12.txt Legates & Willmott December Corrected Precipitation (std. dev.)
285. lwerr00.txt Legates & Willmott Annual Standard Error (mm/year)
286. lwerr01.txt Legates & Willmott January Standard Error (mm/month)
287. lwerr02.txt Legates & Willmott February Standard Error (mm/month)
288. lwerr03.txt Legates & Willmott March Standard Error (mm/month)
289. lwerr04.txt Legates & Willmott April Standard Error (mm/month)
290. lwerr05.txt Legates & Willmott May Standard Error (mm/month)
291. lwerr06.txt Legates & Willmott June Standard Error (mm/month)
292. lwerr07.txt Legates & Willmott July Standard Error (mm/month)
293. lwerr08.txt Legates & Willmott August Standard Error (mm/month)
294. lwerr09.txt Legates & Willmott September Standard Error (mm/month)
295. lwerr10.txt Legates & Willmott October Standard Error (mm/month)
296. lwerr11.txt Legates & Willmott November Standard Error (mm/month)
297. lwerr12.txt Legates & Willmott December Standard Error (mm/month)
298. lwmpr00.txt Legates & Willmott Annual Measured Precipitation (mm/year)
299. lwmpr01.txt Legates & Willmott January Measured Precipitation (mm/month)
300. lwmpr02.txt Legates & Willmott February Measured Precipitation (mm/month)
301. lwmpr03.txt Legates & Willmott March Measured Precipitation (mm/month)

302. lwmpr04.txt Legates & Willmott April Measured Precipitation (mm/month)
303. lwmpr05.txt Legates & Willmott May Measured Precipitation (mm/month)
304. lwmpr06.txt Legates & Willmott June Measured Precipitation (mm/month)
305. lwmpr07.txt Legates & Willmott July Measured Precipitation (mm/month)
306. lwmpr08.txt Legates & Willmott August Measured Precipitation (mm/month)
307. lwmpr09.txt Legates & Willmott September Measured Precipitation (mm/month)
308. lwmpr10.txt Legates & Willmott October Measured Precipitation (mm/month)
309. lwmpr11.txt Legates & Willmott November Measured Precipitation (mm/month)
310. lwmpr12.txt Legates & Willmott December Measured Precipitation (mm/month)
311. lwmsd00.txt Legates & Willmott Annual Measured Precipitation (std. dev.)
312. lwmsd01.txt Legates & Willmott January Measured Precipitation (std. dev.)
313. lwmsd02.txt Legates & Willmott February Measured Precipitation (std. dev.)
314. lwmsd03.txt Legates & Willmott March Measured Precipitation (std. dev.)
315. lwmsd04.txt Legates & Willmott April Measured Precipitation (std. dev.)
316. lwmsd05.txt Legates & Willmott May Measured Precipitation (std. dev.)
317. lwmsd06.txt Legates & Willmott June Measured Precipitation (std. dev.)
318. lwmsd07.txt Legates & Willmott July Measured Precipitation (std. dev.)
319. lwmsd08.txt Legates & Willmott August Measured Precipitation (std. dev.)
320. lwmsd09.txt Legates & Willmott September Measured Precipitation (std. dev.)
321. lwmsd10.txt Legates & Willmott October Measured Precipitation (std. dev.)
322. lwmsd11.txt Legates & Willmott November Measured Precipitation (std. dev.)
323. lwmsd12.txt Legates & Willmott December Measured Precipitation (std. dev.)
324. lwtmp00.txt Legates & Willmott Annual Temperature (0.1C)
325. lwtmp01.txt Legates & Willmott January Temperature (0.1C)
326. lwtmp02.txt Legates & Willmott February Temperature (0.1C)
327. lwtmp03.txt Legates & Willmott March Temperature (0.1C)
328. lwtmp04.txt Legates & Willmott April Temperature (0.1C)
329. lwtmp05.txt Legates & Willmott May Temperature (0.1C)
330. lwtmp06.txt Legates & Willmott June Temperature (0.1C)
331. lwtmp07.txt Legates & Willmott July Temperature (0.1C)
332. lwtmp08.txt Legates & Willmott August Temperature (0.1C)
333. lwtmp09.txt Legates & Willmott September Temperature (0.1C)
334. lwtmp10.txt Legates & Willmott October Temperature (0.1C)
335. lwtmp11.txt Legates & Willmott November Temperature (0.1C)
336. lwtmp12.txt Legates & Willmott December Temperature (0.1C)
337. lwtsd00.txt Legates & Willmott Annual Temperature (std. dev.)
338. lwtsd01.txt Legates & Willmott January Temperature (std. dev.)
339. lwtsd02.txt Legates & Willmott February Temperature (std. dev.)
340. lwtsd03.txt Legates & Willmott March Temperature (std. dev.)
341. lwtsd04.txt Legates & Willmott April Temperature (std. dev.)
342. lwtsd05.txt Legates & Willmott May Temperature (std. dev.)
343. lwtsd06.txt Legates & Willmott June Temperature (std. dev.)
344. lwtsd07.txt Legates & Willmott July Temperature (std. dev.)
345. lwtsd08.txt Legates & Willmott August Temperature (std. dev.)
346. lwtsd09.txt Legates & Willmott September Temperature (std. dev.)
347. lwtsd10.txt Legates & Willmott October Temperature (std. dev.)
348. lwtsd11.txt Legates & Willmott November Temperature (std. dev.)
349. lwtsd12.txt Legates & Willmott December Temperature (std. dev.)
350. macult.txt Matthews Cultivation Intensity
351. malbfa.txt Matthews Fall Albedo (% X 100)
352. malbsm.txt Matthews Summer Albedo (% X 100)
353. malbsp.txt Matthews Spring Albedo (% X 100)
354. malbwn.txt Matthews Winter Albedo (% X 100)
355. maveg.txt Matthews Vegetation Types

Niche Modeling

356. mev8504.txt April 1985 Experimental Vegetation Index
357. mev8505.txt May 1985 Experimental Vegetation Index
358. mev8506.txt June 1985 Experimental Vegetation Index
359. mev8507.txt July 1985 Experimental Vegetation Index
360. mev8508.txt August 1985 Experimental Vegetation Index
361. mev8509.txt September 1985 Experimental Vegetation Index
362. mev8510.txt October 1985 Experimental Vegetation Index
363. mev8511.txt November 1985 Experimental Vegetation Index
364. mev8512.txt December 1985 Experimental Vegetation Index
365. mev8601.txt January 1986 Experimental Vegetation Index
366. mev8602.txt February 1986 Experimental Vegetation Index
367. mev8603.txt March 1986 Experimental Vegetation Index
368. mev8604.txt April 1986 Experimental Vegetation Index
369. mev8605.txt May 1986 Experimental Vegetation Index
370. mev8606.txt June 1986 Experimental Vegetation Index
371. mev8607.txt July 1986 Experimental Vegetation Index
372. mev8608.txt August 1986 Experimental Vegetation Index
373. mev8609.txt September 1986 Experimental Vegetation Index
374. mev8610.txt October 1986 Experimental Vegetation Index
375. mev8611.txt November 1986 Experimental Vegetation Index
376. mev8612.txt December 1986 Experimental Vegetation Index
377. mev8701.txt January 1987 Experimental Vegetation Index
378. mev8702.txt February 1987 Experimental Vegetation Index
379. mev8703.txt March 1987 Experimental Vegetation Index
380. mev8704.txt April 1987 Experimental Vegetation Index
381. mev8705.txt May 1987 Experimental Vegetation Index
382. mev8706.txt June 1987 Experimental Vegetation Index
383. mev8707.txt July 1987 Experimental Vegetation Index
384. mev8708.txt August 1987 Experimental Vegetation Index
385. mev8709.txt September 1987 Experimental Vegetation Index
386. mev8710.txt October 1987 Experimental Vegetation Index
387. mev8711.txt November 1987 Experimental Vegetation Index
388. mev8712.txt December 1987 Experimental Vegetation Index
389. mev8801.txt January 1988 Experimental Vegetation Index
390. mev8802.txt February 1988 Experimental Vegetation Index
391. mev8803.txt March 1988 Experimental Vegetation Index
392. mev8804.txt April 1988 Experimental Vegetation Index
393. mev8805.txt May 1988 Experimental Vegetation Index
394. mev8806.txt June 1988 Experimental Vegetation Index
395. mev8807.txt July 1988 Experimental Vegetation Index
396. mev8808.txt August 1988 Experimental Vegetation Index
397. mev8809.txt September 1988 Experimental Vegetation Index
398. mev8810.txt October 1988 Experimental Vegetation Index
399. mev8811.txt November 1988 Experimental Vegetation Index
400. mev8812.txt December 1988 Experimental Vegetation Index
401. mev8901.txt January 1989 Experimental Vegetation Index
402. mev8902.txt February 1989 Experimental Vegetation Index
403. mev8903.txt March 1989 Experimental Vegetation Index
404. mev8904.txt April 1989 Experimental Vegetation Index
405. mev8905.txt May 1989 Experimental Vegetation Index
406. mev8906.txt June 1989 Experimental Vegetation Index
407. mev8907.txt July 1989 Experimental Vegetation Index
408. mev8908.txt August 1989 Experimental Vegetation Index
409. mev8909.txt September 1989 Experimental Vegetation Index

410. mev8910.txt October 1989 Experimental Vegetation Index
411. mev8911.txt November 1989 Experimental Vegetation Index
412. mev8912.txt December 1989 Experimental Vegetation Index
413. mev9001.txt January 1990 Experimental Vegetation Index
414. mev9002.txt February 1990 Experimental Vegetation Index
415. mev9003.txt March 1990 Experimental Vegetation Index
416. mev9004.txt April 1990 Experimental Vegetation Index
417. mev9005.txt May 1990 Experimental Vegetation Index
418. mev9006.txt June 1990 Experimental Vegetation Index
419. mev9007.txt July 1990 Experimental Vegetation Index
420. mev9008.txt August 1990 Experimental Vegetation Index
421. mev9009.txt September 1990 Experimental Vegetation Index
422. mev9010.txt October 1990 Experimental Vegetation Index
423. mev9011.txt November 1990 Experimental Vegetation Index
424. mev9012.txt December 1990 Experimental Vegetation Index
425. mfwfrin.txt Matthews and Fung Fractional Inundation
426. mfwsol.txt FAO Soil Types of Matthews & Fung Wetland Locations
427. mfwsrc.txt Matthews and Fung Wetland Data Source
428. mfwveg.txt Matthews and Fung Vegetation Type
429. mfwwet.txt Matthews and Fung Wetland Type
430. mgv0001.txt Average January Generalized Global Vegetation Index
431. mgv0002.txt Average February Generalized Global Vegetation Index
432. mgv0003.txt Average March Generalized Global Vegetation Index
433. mgv0004.txt Average April Generalized Global Vegetation Index
434. mgv0005.txt Average May Generalized Global Vegetation Index
435. mgv0006.txt Average June Generalized Global Vegetation Index
436. mgv0007.txt Average July Generalized Global Vegetation Index
437. mgv0008.txt Average August Generalized Global Vegetation Index
438. mgv0009.txt Average September Generalized Global Vegetation Index
439. mgv0010.txt Average October Generalized Global Vegetation Index
440. mgv0011.txt Average November Generalized Global Vegetation Index
441. mgv0012.txt Average December Generalized Global Vegetation Index
442. mgv8504.txt April 1985 Generalized Global Vegetation Index
443. mgv8505.txt May 1985 Generalized Global Vegetation Index
444. mgv8506.txt June 1985 Generalized Global Vegetation Index
445. mgv8507.txt July 1985 Generalized Global Vegetation Index
446. mgv8508.txt August 1985 Generalized Global Vegetation Index
447. mgv8509.txt September 1985 Generalized Global Vegetation Index
448. mgv8510.txt October 1985 Generalized Global Vegetation Index
449. mgv8511.txt November 1985 Generalized Global Vegetation Index
450. mgv8512.txt December 1985 Generalized Global Vegetation Index
451. mgv8601.txt January 1986 Generalized Global Vegetation Index
452. mgv8602.txt February 1986 Generalized Global Vegetation Index
453. mgv8603.txt March 1986 Generalized Global Vegetation Index
454. mgv8604.txt April 1986 Generalized Global Vegetation Index
455. mgv8605.txt May 1986 Generalized Global Vegetation Index
456. mgv8606.txt June 1986 Generalized Global Vegetation Index
457. mgv8607.txt July 1986 Generalized Global Vegetation Index
458. mgv8608.txt August 1986 Generalized Global Vegetation Index
459. mgv8609.txt September 1986 Generalized Global Vegetation Index
460. mgv8610.txt October 1986 Generalized Global Vegetation Index
461. mgv8611.txt November 1986 Generalized Global Vegetation Index
462. mgv8612.txt December 1986 Generalized Global Vegetation Index
463. mgv8701.txt January 1987 Generalized Global Vegetation Index

464. mgv8702.txt February 1987 Generalized Global Vegetation Index
465. mgv8703.txt March 1987 Generalized Global Vegetation Index
466. mgv8704.txt April 1987 Generalized Global Vegetation Index
467. mgv8705.txt May 1987 Generalized Global Vegetation Index
468. mgv8706.txt June 1987 Generalized Global Vegetation Index
469. mgv8707.txt July 1987 Generalized Global Vegetation Index
470. mgv8708.txt August 1987 Generalized Global Vegetation Index
471. mgv8709.txt September 1987 Generalized Global Vegetation Index
472. mgv8710.txt October 1987 Generalized Global Vegetation Index
473. mgv8711.txt November 1987 Generalized Global Vegetation Index
474. mgv8712.txt December 1987 Generalized Global Vegetation Index
475. mgv8801.txt January 1988 Generalized Global Vegetation Index
476. mgv8802.txt February 1988 Generalized Global Vegetation Index
477. mgv8803.txt March 1988 Generalized Global Vegetation Index
478. mgv8804.txt April 1988 Generalized Global Vegetation Index
479. mgv8805.txt May 1988 Generalized Global Vegetation Index
480. mgv8806.txt June 1988 Generalized Global Vegetation Index
481. mgv8807.txt July 1988 Generalized Global Vegetation Index
482. mgv8808.txt August 1988 Generalized Global Vegetation Index
483. mgv8809.txt September 1988 Generalized Global Vegetation Index
484. mgv8810.txt October 1988 Generalized Global Vegetation Index
485. mgv8811.txt November 1988 Generalized Global Vegetation Index
486. mgv8812.txt December 1988 Generalized Global Vegetation Index
487. mgvc186.txt 1986 MGV PCA Component 1
488. mgvc187.txt 1987 MGV PCA Component 1
489. mgvc188.txt 1988 MGV PCA Component 1
490. mgvc286.txt 1986 MGV PCA Component 2
491. mgvc287.txt 1987 MGV PCA Component 2
492. mgvc288.txt 1988 MGV PCA Component 2
493. mgvc386.txt 1986 MGV PCA Component 3
494. mgvc387.txt 1987 MGV PCA Component 3
495. mgvc388.txt 1988 MGV PCA Component 3
496. mgvc486.txt 1986 MGV PCA Component 4
497. mgvc487.txt 1987 MGV PCA Component 4
498. mgvc488.txt 1988 MGV PCA Component 4
499. mwcoast.dvc Micro World Data Bank II Coasts
500. mwisland.dvc Micro World Data Bank II Islands
501. mwlake.dvc Micro World Data Bank II Lakes
502. mwnation.dvc Micro World Data Bank II Countries
503. mwriver.dvc Micro World Data Bank II Rivers
504. mwstate.dvc Micro World Data Bank II States
505. n00an1.1.txt World Ocean Atlas 2001 1 degree annual analyzed mean - nitrate at depth 0 metre
506. n00an1.10.txt World Ocean Atlas 2001 1 degree annual analyzed mean - nitrate at depth 200 metre
507. n00an1.11.txt World Ocean Atlas 2001 1 degree annual analyzed mean - nitrate at depth 250 metre
508. n00an1.12.txt World Ocean Atlas 2001 1 degree annual analyzed mean - nitrate at depth 300 metre
509. n00an1.13.txt World Ocean Atlas 2001 1 degree annual analyzed mean - nitrate at depth 400 metre
510. n00an1.14.txt World Ocean Atlas 2001 1 degree annual analyzed mean - nitrate at depth 500 metre
511. n00an1.15.txt World Ocean Atlas 2001 1 degree annual analyzed mean - nitrate at depth 600 metre
512. n00an1.16.txt World Ocean Atlas 2001 1 degree annual analyzed mean - nitrate at depth 700 metre
513. n00an1.17.txt World Ocean Atlas 2001 1 degree annual analyzed mean - nitrate at depth 800 metre
514. n00an1.18.txt World Ocean Atlas 2001 1 degree annual analyzed mean - nitrate at depth 900 metre
515. n00an1.19.txt World Ocean Atlas 2001 1 degree annual analyzed mean - nitrate at depth 1000 metre
516. n00an1.2.txt World Ocean Atlas 2001 1 degree annual analyzed mean - nitrate at depth 10 metre
517. n00an1.20.txt World Ocean Atlas 2001 1 degree annual analyzed mean - nitrate at depth 1100 metre

518. n00an1.21.txt World Ocean Atlas 2001 1 degree annual analyzed mean - nitrate at depth 1200 metre
519. n00an1.22.txt World Ocean Atlas 2001 1 degree annual analyzed mean - nitrate at depth 1300 metre
520. n00an1.23.txt World Ocean Atlas 2001 1 degree annual analyzed mean - nitrate at depth 1400 metre
521. n00an1.24.txt World Ocean Atlas 2001 1 degree annual analyzed mean - nitrate at depth 1500 metre
522. n00an1.25.txt World Ocean Atlas 2001 1 degree annual analyzed mean - nitrate at depth 1750 metre
523. n00an1.26.txt World Ocean Atlas 2001 1 degree annual analyzed mean - nitrate at depth 2000 metre
524. n00an1.27.txt World Ocean Atlas 2001 1 degree annual analyzed mean - nitrate at depth 2500 metre
525. n00an1.28.txt World Ocean Atlas 2001 1 degree annual analyzed mean - nitrate at depth 3000 metre
526. n00an1.29.txt World Ocean Atlas 2001 1 degree annual analyzed mean - nitrate at depth 3500 metre
527. n00an1.3.txt World Ocean Atlas 2001 1 degree annual analyzed mean - nitrate at depth 20 metre
528. n00an1.30.txt World Ocean Atlas 2001 1 degree annual analyzed mean - nitrate at depth 4000 metre
529. n00an1.31.txt World Ocean Atlas 2001 1 degree annual analyzed mean - nitrate at depth 4500 metre
530. n00an1.32.txt World Ocean Atlas 2001 1 degree annual analyzed mean - nitrate at depth 5000 metre
531. n00an1.33.txt World Ocean Atlas 2001 1 degree annual analyzed mean - nitrate at depth 5500 metre
532. n00an1.4.txt World Ocean Atlas 2001 1 degree annual analyzed mean - nitrate at depth 30 metre
533. n00an1.5.txt World Ocean Atlas 2001 1 degree annual analyzed mean - nitrate at depth 50 metre
534. n00an1.6.txt World Ocean Atlas 2001 1 degree annual analyzed mean - nitrate at depth 75 metre
535. n00an1.7.txt World Ocean Atlas 2001 1 degree annual analyzed mean - nitrate at depth 100 metre
536. n00an1.8.txt World Ocean Atlas 2001 1 degree annual analyzed mean - nitrate at depth 125 metre
537. n00an1.9.txt World Ocean Atlas 2001 1 degree annual analyzed mean - nitrate at depth 150 metre
538. n00sd1.1.txt World Ocean Atlas 2001 1 degree standard deviation - nitrate at depth 0 metre
539. n00sd1.10.txt World Ocean Atlas 2001 1 degree standard deviation - nitrate at depth 200 metre
540. n00sd1.11.txt World Ocean Atlas 2001 1 degree standard deviation - nitrate at depth 250 metre
541. n00sd1.12.txt World Ocean Atlas 2001 1 degree standard deviation - nitrate at depth 300 metre
542. n00sd1.13.txt World Ocean Atlas 2001 1 degree standard deviation - nitrate at depth 400 metre
543. n00sd1.14.txt World Ocean Atlas 2001 1 degree standard deviation - nitrate at depth 500 metre
544. n00sd1.15.txt World Ocean Atlas 2001 1 degree standard deviation - nitrate at depth 600 metre
545. n00sd1.16.txt World Ocean Atlas 2001 1 degree standard deviation - nitrate at depth 700 metre
546. n00sd1.17.txt World Ocean Atlas 2001 1 degree standard deviation - nitrate at depth 800 metre
547. n00sd1.18.txt World Ocean Atlas 2001 1 degree standard deviation - nitrate at depth 900 metre
548. n00sd1.19.txt World Ocean Atlas 2001 1 degree standard deviation - nitrate at depth 1000 metre
549. n00sd1.2.txt World Ocean Atlas 2001 1 degree standard deviation - nitrate at depth 10 metre
550. n00sd1.20.txt World Ocean Atlas 2001 1 degree standard deviation - nitrate at depth 1100 metre
551. n00sd1.21.txt World Ocean Atlas 2001 1 degree standard deviation - nitrate at depth 1200 metre
552. n00sd1.22.txt World Ocean Atlas 2001 1 degree standard deviation - nitrate at depth 1300 metre
553. n00sd1.23.txt World Ocean Atlas 2001 1 degree standard deviation - nitrate at depth 1400 metre
554. n00sd1.24.txt World Ocean Atlas 2001 1 degree standard deviation - nitrate at depth 1500 metre
555. n00sd1.25.txt World Ocean Atlas 2001 1 degree standard deviation - nitrate at depth 1750 metre
556. n00sd1.26.txt World Ocean Atlas 2001 1 degree standard deviation - nitrate at depth 2000 metre
557. n00sd1.27.txt World Ocean Atlas 2001 1 degree standard deviation - nitrate at depth 2500 metre
558. n00sd1.28.txt World Ocean Atlas 2001 1 degree standard deviation - nitrate at depth 3000 metre
559. n00sd1.29.txt World Ocean Atlas 2001 1 degree standard deviation - nitrate at depth 3500 metre
560. n00sd1.3.txt World Ocean Atlas 2001 1 degree standard deviation - nitrate at depth 20 metre
561. n00sd1.30.txt World Ocean Atlas 2001 1 degree standard deviation - nitrate at depth 4000 metre
562. n00sd1.31.txt World Ocean Atlas 2001 1 degree standard deviation - nitrate at depth 4500 metre
563. n00sd1.32.txt World Ocean Atlas 2001 1 degree standard deviation - nitrate at depth 5000 metre
564. n00sd1.33.txt World Ocean Atlas 2001 1 degree standard deviation - nitrate at depth 5500 metre
565. n00sd1.4.txt World Ocean Atlas 2001 1 degree standard deviation - nitrate at depth 30 metre
566. n00sd1.5.txt World Ocean Atlas 2001 1 degree standard deviation - nitrate at depth 50 metre
567. n00sd1.6.txt World Ocean Atlas 2001 1 degree standard deviation - nitrate at depth 75 metre
568. n00sd1.7.txt World Ocean Atlas 2001 1 degree standard deviation - nitrate at depth 100 metre
569. n00sd1.8.txt World Ocean Atlas 2001 1 degree standard deviation - nitrate at depth 125 metre
570. n00sd1.9.txt World Ocean Atlas 2001 1 degree standard deviation - nitrate at depth 150 metre
571. o00an1.1.txt World Ocean Atlas 2001 1 degree annual analyzed mean - dissovd oxygen at depth 0 metre

572. o00an1.10.txt World Ocean Atlas 2001 1 degree annual analyzed mean - dissovd oxygen at depth 200 metre

573. o00an1.11.txt World Ocean Atlas 2001 1 degree annual analyzed mean - dissovd oxygen at depth 250 metre

574. o00an1.12.txt World Ocean Atlas 2001 1 degree annual analyzed mean - dissovd oxygen at depth 300 metre

575. o00an1.13.txt World Ocean Atlas 2001 1 degree annual analyzed mean - dissovd oxygen at depth 400 metre

576. o00an1.14.txt World Ocean Atlas 2001 1 degree annual analyzed mean - dissovd oxygen at depth 500 metre

577. o00an1.15.txt World Ocean Atlas 2001 1 degree annual analyzed mean - dissovd oxygen at depth 600 metre

578. o00an1.16.txt World Ocean Atlas 2001 1 degree annual analyzed mean - dissovd oxygen at depth 700 metre

579. o00an1.17.txt World Ocean Atlas 2001 1 degree annual analyzed mean - dissovd oxygen at depth 800 metre

580. o00an1.18.txt World Ocean Atlas 2001 1 degree annual analyzed mean - dissovd oxygen at depth 900 metre

581. o00an1.19.txt World Ocean Atlas 2001 1 degree annual analyzed mean - dissovd oxygen at depth 1000 metre

582. o00an1.2.txt World Ocean Atlas 2001 1 degree annual analyzed mean - dissovd oxygen at depth 10 metre

583. o00an1.20.txt World Ocean Atlas 2001 1 degree annual analyzed mean - dissovd oxygen at depth 1100 metre

584. o00an1.21.txt World Ocean Atlas 2001 1 degree annual analyzed mean - dissovd oxygen at depth 1200 metre

585. o00an1.22.txt World Ocean Atlas 2001 1 degree annual analyzed mean - dissovd oxygen at depth 1300 metre

586. o00an1.23.txt World Ocean Atlas 2001 1 degree annual analyzed mean - dissovd oxygen at depth 1400 metre

587. o00an1.24.txt World Ocean Atlas 2001 1 degree annual analyzed mean - dissovd oxygen at depth 1500 metre

588. o00an1.25.txt World Ocean Atlas 2001 1 degree annual analyzed mean - dissovd oxygen at depth 1750 metre

589. o00an1.26.txt World Ocean Atlas 2001 1 degree annual analyzed mean - dissovd oxygen at depth 2000 metre

590. o00an1.27.txt World Ocean Atlas 2001 1 degree annual analyzed mean - dissovd oxygen at depth 2500 metre

591. o00an1.28.txt World Ocean Atlas 2001 1 degree annual analyzed mean - dissovd oxygen at depth 3000 metre

592. o00an1.29.txt World Ocean Atlas 2001 1 degree annual analyzed mean - dissovd oxygen at depth 3500 metre

593. o00an1.3.txt World Ocean Atlas 2001 1 degree annual analyzed mean - dissovd oxygen at depth 20 metre

594. o00an1.30.txt World Ocean Atlas 2001 1 degree annual analyzed mean - dissovd oxygen at depth 4000 metre

595. o00an1.31.txt World Ocean Atlas 2001 1 degree annual analyzed mean - dissovd oxygen at depth 4500 metre

596. o00an1.32.txt World Ocean Atlas 2001 1 degree annual analyzed mean - dissovd oxygen at depth 5000 metre

597. o00an1.33.txt World Ocean Atlas 2001 1 degree annual analyzed mean - dissovd oxygen at depth 5500 metre

598. o00an1.4.txt World Ocean Atlas 2001 1 degree annual analyzed mean - dissovd oxygen at depth 30 metre

599. o00an1.5.txt World Ocean Atlas 2001 1 degree annual analyzed mean - dissovd oxygen at depth 50 metre

600. o00an1.6.txt World Ocean Atlas 2001 1 degree annual analyzed mean - dissovd oxygen at depth 75 metre

601. o00an1.7.txt World Ocean Atlas 2001 1 degree annual analyzed mean - dissovd oxygen at depth 100 metre

602. o00an1.8.txt World Ocean Atlas 2001 1 degree annual analyzed mean - dissovd oxygen at depth 125 metre

603. o00an1.9.txt World Ocean Atlas 2001 1 degree annual analyzed mean - dissovd oxygen at depth 150 metre

604. o00sd1.1.txt World Ocean Atlas 2001 1 degree standard deviation - dissovd oxygen at depth 0 metre

605. o00sd1.10.txt World Ocean Atlas 2001 1 degree standard deviation - dissovd oxygen at depth 200 metre

606. o00sd1.11.txt World Ocean Atlas 2001 1 degree standard deviation - dissovd oxygen at depth 250 metre

607. o00sd1.12.txt World Ocean Atlas 2001 1 degree standard deviation - dissovd oxygen at depth 300 metre

608. o00sd1.13.txt World Ocean Atlas 2001 1 degree standard deviation - dissovd oxygen at depth 400 metre

609. o00sd1.14.txt World Ocean Atlas 2001 1 degree standard deviation - dissovd oxygen at depth 500 metre

610. o00sd1.15.txt World Ocean Atlas 2001 1 degree standard deviation - dissovd oxygen at depth 600 metre
611. o00sd1.16.txt World Ocean Atlas 2001 1 degree standard deviation - dissovd oxygen at depth 700 metre
612. o00sd1.17.txt World Ocean Atlas 2001 1 degree standard deviation - dissovd oxygen at depth 800 metre
613. o00sd1.18.txt World Ocean Atlas 2001 1 degree standard deviation - dissovd oxygen at depth 900 metre
614. o00sd1.19.txt World Ocean Atlas 2001 1 degree standard deviation - dissovd oxygen at depth 1000 metre
615. o00sd1.2.txt World Ocean Atlas 2001 1 degree standard deviation - dissovd oxygen at depth 10 metre
616. o00sd1.20.txt World Ocean Atlas 2001 1 degree standard deviation - dissovd oxygen at depth 1100 metre
617. o00sd1.21.txt World Ocean Atlas 2001 1 degree standard deviation - dissovd oxygen at depth 1200 metre
618. o00sd1.22.txt World Ocean Atlas 2001 1 degree standard deviation - dissovd oxygen at depth 1300 metre
619. o00sd1.23.txt World Ocean Atlas 2001 1 degree standard deviation - dissovd oxygen at depth 1400 metre
620. o00sd1.24.txt World Ocean Atlas 2001 1 degree standard deviation - dissovd oxygen at depth 1500 metre
621. o00sd1.25.txt World Ocean Atlas 2001 1 degree standard deviation - dissovd oxygen at depth 1750 metre
622. o00sd1.26.txt World Ocean Atlas 2001 1 degree standard deviation - dissovd oxygen at depth 2000 metre
623. o00sd1.27.txt World Ocean Atlas 2001 1 degree standard deviation - dissovd oxygen at depth 2500 metre
624. o00sd1.28.txt World Ocean Atlas 2001 1 degree standard deviation - dissovd oxygen at depth 3000 metre
625. o00sd1.29.txt World Ocean Atlas 2001 1 degree standard deviation - dissovd oxygen at depth 3500 metre
626. o00sd1.3.txt World Ocean Atlas 2001 1 degree standard deviation - dissovd oxygen at depth 20 metre
627. o00sd1.30.txt World Ocean Atlas 2001 1 degree standard deviation - dissovd oxygen at depth 4000 metre
628. o00sd1.31.txt World Ocean Atlas 2001 1 degree standard deviation - dissovd oxygen at depth 4500 metre
629. o00sd1.32.txt World Ocean Atlas 2001 1 degree standard deviation - dissovd oxygen at depth 5000 metre
630. o00sd1.33.txt World Ocean Atlas 2001 1 degree standard deviation - dissovd oxygen at depth 5500 metre
631. o00sd1.4.txt World Ocean Atlas 2001 1 degree standard deviation - dissovd oxygen at depth 30 metre
632. o00sd1.5.txt World Ocean Atlas 2001 1 degree standard deviation - dissovd oxygen at depth 50 metre
633. o00sd1.6.txt World Ocean Atlas 2001 1 degree standard deviation - dissovd oxygen at depth 75 metre
634. o00sd1.7.txt World Ocean Atlas 2001 1 degree standard deviation - dissovd oxygen at depth 100 metre
635. o00sd1.8.txt World Ocean Atlas 2001 1 degree standard deviation - dissovd oxygen at depth 125 metre
636. o00sd1.9.txt World Ocean Atlas 2001 1 degree standard deviation - dissovd oxygen at depth 150 metre
637. owe13a.txt Olson World Ecosystems Version 1.3A
638. owe14d.txt Olson World Ecosystem Classes Version 1.4D
639. owe14dr.txt Resolution codes for OWE1.4D
640. p00an1.1.txt World Ocean Atlas 2001 1 degree annual analyzed mean - phosphate at depth 0 metre
641. p00an1.10.txt World Ocean Atlas 2001 1 degree annual analyzed mean - phosphate at depth 200 metre
642. p00an1.11.txt World Ocean Atlas 2001 1 degree annual analyzed mean - phosphate at depth 250 metre
643. p00an1.12.txt World Ocean Atlas 2001 1 degree annual analyzed mean - phosphate at depth 300 metre
644. p00an1.13.txt World Ocean Atlas 2001 1 degree annual analyzed mean - phosphate at depth 400 metre
645. p00an1.14.txt World Ocean Atlas 2001 1 degree annual analyzed mean - phosphate at depth 500 metre
646. p00an1.15.txt World Ocean Atlas 2001 1 degree annual analyzed mean - phosphate at depth 600 metre
647. p00an1.16.txt World Ocean Atlas 2001 1 degree annual analyzed mean - phosphate at depth 700 metre
648. p00an1.17.txt World Ocean Atlas 2001 1 degree annual analyzed mean - phosphate at depth 800 metre
649. p00an1.18.txt World Ocean Atlas 2001 1 degree annual analyzed mean - phosphate at depth 900 metre
650. p00an1.19.txt World Ocean Atlas 2001 1 degree annual analyzed mean - phosphate at depth 1000 metre
651. p00an1.2.txt World Ocean Atlas 2001 1 degree annual analyzed mean - phosphate at depth 10 metre
652. p00an1.20.txt World Ocean Atlas 2001 1 degree annual analyzed mean - phosphate at depth 1100 metre
653. p00an1.21.txt World Ocean Atlas 2001 1 degree annual analyzed mean - phosphate at depth 1200 metre
654. p00an1.22.txt World Ocean Atlas 2001 1 degree annual analyzed mean - phosphate at depth 1300 metre
655. p00an1.23.txt World Ocean Atlas 2001 1 degree annual analyzed mean - phosphate at depth 1400 metre
656. p00an1.24.txt World Ocean Atlas 2001 1 degree annual analyzed mean - phosphate at depth 1500 metre
657. p00an1.25.txt World Ocean Atlas 2001 1 degree annual analyzed mean - phosphate at depth 1750 metre
658. p00an1.26.txt World Ocean Atlas 2001 1 degree annual analyzed mean - phosphate at depth 2000 metre
659. p00an1.27.txt World Ocean Atlas 2001 1 degree annual analyzed mean - phosphate at depth 2500 metre
660. p00an1.28.txt World Ocean Atlas 2001 1 degree annual analyzed mean - phosphate at depth 3000 metre
661. p00an1.29.txt World Ocean Atlas 2001 1 degree annual analyzed mean - phosphate at depth 3500 metre
662. p00an1.3.txt World Ocean Atlas 2001 1 degree annual analyzed mean - phosphate at depth 20 metre
663. p00an1.30.txt World Ocean Atlas 2001 1 degree annual analyzed mean - phosphate at depth 4000 metre

664. p00an1.31.txt World Ocean Atlas 2001 1 degree annual analyzed mean - phosphate at depth 4500 metre
665. p00an1.32.txt World Ocean Atlas 2001 1 degree annual analyzed mean - phosphate at depth 5000 metre
666. p00an1.33.txt World Ocean Atlas 2001 1 degree annual analyzed mean - phosphate at depth 5500 metre
667. p00an1.4.txt World Ocean Atlas 2001 1 degree annual analyzed mean - phosphate at depth 30 metre
668. p00an1.5.txt World Ocean Atlas 2001 1 degree annual analyzed mean - phosphate at depth 50 metre
669. p00an1.6.txt World Ocean Atlas 2001 1 degree annual analyzed mean - phosphate at depth 75 metre
670. p00an1.7.txt World Ocean Atlas 2001 1 degree annual analyzed mean - phosphate at depth 100 metre
671. p00an1.8.txt World Ocean Atlas 2001 1 degree annual analyzed mean - phosphate at depth 125 metre
672. p00an1.9.txt World Ocean Atlas 2001 1 degree annual analyzed mean - phosphate at depth 150 metre
673. p00sd1.1.txt World Ocean Atlas 2001 1 degree standard deviation - phosphate at depth 0 metre
674. p00sd1.10.txt World Ocean Atlas 2001 1 degree standard deviation - phosphate at depth 200 metre
675. p00sd1.11.txt World Ocean Atlas 2001 1 degree standard deviation - phosphate at depth 250 metre
676. p00sd1.12.txt World Ocean Atlas 2001 1 degree standard deviation - phosphate at depth 300 metre
677. p00sd1.13.txt World Ocean Atlas 2001 1 degree standard deviation - phosphate at depth 400 metre
678. p00sd1.14.txt World Ocean Atlas 2001 1 degree standard deviation - phosphate at depth 500 metre
679. p00sd1.15.txt World Ocean Atlas 2001 1 degree standard deviation - phosphate at depth 600 metre
680. p00sd1.16.txt World Ocean Atlas 2001 1 degree standard deviation - phosphate at depth 700 metre
681. p00sd1.17.txt World Ocean Atlas 2001 1 degree standard deviation - phosphate at depth 800 metre
682. p00sd1.18.txt World Ocean Atlas 2001 1 degree standard deviation - phosphate at depth 900 metre
683. p00sd1.19.txt World Ocean Atlas 2001 1 degree standard deviation - phosphate at depth 1000 metre
684. p00sd1.2.txt World Ocean Atlas 2001 1 degree standard deviation - phosphate at depth 10 metre
685. p00sd1.20.txt World Ocean Atlas 2001 1 degree standard deviation - phosphate at depth 1100 metre
686. p00sd1.21.txt World Ocean Atlas 2001 1 degree standard deviation - phosphate at depth 1200 metre
687. p00sd1.22.txt World Ocean Atlas 2001 1 degree standard deviation - phosphate at depth 1300 metre
688. p00sd1.23.txt World Ocean Atlas 2001 1 degree standard deviation - phosphate at depth 1400 metre
689. p00sd1.24.txt World Ocean Atlas 2001 1 degree standard deviation - phosphate at depth 1500 metre
690. p00sd1.25.txt World Ocean Atlas 2001 1 degree standard deviation - phosphate at depth 1750 metre
691. p00sd1.26.txt World Ocean Atlas 2001 1 degree standard deviation - phosphate at depth 2000 metre
692. p00sd1.27.txt World Ocean Atlas 2001 1 degree standard deviation - phosphate at depth 2500 metre
693. p00sd1.28.txt World Ocean Atlas 2001 1 degree standard deviation - phosphate at depth 3000 metre
694. p00sd1.29.txt World Ocean Atlas 2001 1 degree standard deviation - phosphate at depth 3500 metre
695. p00sd1.3.txt World Ocean Atlas 2001 1 degree standard deviation - phosphate at depth 20 metre
696. p00sd1.30.txt World Ocean Atlas 2001 1 degree standard deviation - phosphate at depth 4000 metre
697. p00sd1.31.txt World Ocean Atlas 2001 1 degree standard deviation - phosphate at depth 4500 metre
698. p00sd1.32.txt World Ocean Atlas 2001 1 degree standard deviation - phosphate at depth 5000 metre
699. p00sd1.33.txt World Ocean Atlas 2001 1 degree standard deviation - phosphate at depth 5500 metre
700. p00sd1.4.txt World Ocean Atlas 2001 1 degree standard deviation - phosphate at depth 30 metre
701. p00sd1.5.txt World Ocean Atlas 2001 1 degree standard deviation - phosphate at depth 50 metre
702. p00sd1.6.txt World Ocean Atlas 2001 1 degree standard deviation - phosphate at depth 75 metre
703. p00sd1.7.txt World Ocean Atlas 2001 1 degree standard deviation - phosphate at depth 100 metre
704. p00sd1.8.txt World Ocean Atlas 2001 1 degree standard deviation - phosphate at depth 125 metre
705. p00sd1.9.txt World Ocean Atlas 2001 1 degree standard deviation - phosphate at depth 150 metre
706. PP_Period_9809_9908.txt Primary Production in Period 9809 to 9908
707. prec_1.txt Precipitation of January
708. prec_10.txt Precipitation of October
709. prec_10_2.5m.txt prec_10 2.5 minutes
710. prec_11.txt Precipitation of November
711. prec_11_2.5m.txt prec_11 2.5 minutes
712. prec_12.txt Precipitation of December
713. prec_12_2.5m.txt prec_12 2.5 minutes
714. prec_1_2.5m.txt prec_1 2.5 minutes
715. prec_2.txt Precipitation of February
716. prec_2_2.5m.txt prec_2 2.5 minutes
717. prec_3.txt Precipitation of March

718. prec_3_2.5m.txt prec_3 2.5 minutes
719. prec_4.txt Precipitation of April
720. prec_4_2.5m.txt prec_4 2.5 minutes
721. prec_5.txt Precipitation of May
722. prec_5_2.5m.txt prec_5 2.5 minutes
723. prec_6.txt Precipitation of June
724. prec_6_2.5m.txt prec_6 2.5 minutes
725. prec_7.txt Precipitation of July
726. prec_7_2.5m.txt prec_7 2.5 minutes
727. prec_8.txt Precipitation of August
728. prec_8_2.5m.txt prec_8 2.5 minutes
729. prec_9.txt Precipitation of September
730. prec_9_2.5m.txt prec_9 2.5 minutes
731. s00an1.1.txt World Ocean Atlas 2001 1 degree annual analyzed mean - salinity at depth 0 metre
732. s00an1.10.txt World Ocean Atlas 2001 1 degree annual analyzed mean - salinity at depth 200 metre
733. s00an1.11.txt World Ocean Atlas 2001 1 degree annual analyzed mean - salinity at depth 250 metre
734. s00an1.12.txt World Ocean Atlas 2001 1 degree annual analyzed mean - salinity at depth 300 metre
735. s00an1.13.txt World Ocean Atlas 2001 1 degree annual analyzed mean - salinity at depth 400 metre
736. s00an1.14.txt World Ocean Atlas 2001 1 degree annual analyzed mean - salinity at depth 500 metre
737. s00an1.15.txt World Ocean Atlas 2001 1 degree annual analyzed mean - salinity at depth 600 metre
738. s00an1.16.txt World Ocean Atlas 2001 1 degree annual analyzed mean - salinity at depth 700 metre
739. s00an1.17.txt World Ocean Atlas 2001 1 degree annual analyzed mean - salinity at depth 800 metre
740. s00an1.18.txt World Ocean Atlas 2001 1 degree annual analyzed mean - salinity at depth 900 metre
741. s00an1.19.txt World Ocean Atlas 2001 1 degree annual analyzed mean - salinity at depth 1000 metre
742. s00an1.2.txt World Ocean Atlas 2001 1 degree annual analyzed mean - salinity at depth 10 metre
743. s00an1.20.txt World Ocean Atlas 2001 1 degree annual analyzed mean - salinity at depth 1100 metre
744. s00an1.21.txt World Ocean Atlas 2001 1 degree annual analyzed mean - salinity at depth 1200 metre
745. s00an1.22.txt World Ocean Atlas 2001 1 degree annual analyzed mean - salinity at depth 1300 metre
746. s00an1.23.txt World Ocean Atlas 2001 1 degree annual analyzed mean - salinity at depth 1400 metre
747. s00an1.24.txt World Ocean Atlas 2001 1 degree annual analyzed mean - salinity at depth 1500 metre
748. s00an1.25.txt World Ocean Atlas 2001 1 degree annual analyzed mean - salinity at depth 1750 metre
749. s00an1.26.txt World Ocean Atlas 2001 1 degree annual analyzed mean - salinity at depth 2000 metre
750. s00an1.27.txt World Ocean Atlas 2001 1 degree annual analyzed mean - salinity at depth 2500 metre
751. s00an1.28.txt World Ocean Atlas 2001 1 degree annual analyzed mean - salinity at depth 3000 metre
752. s00an1.29.txt World Ocean Atlas 2001 1 degree annual analyzed mean - salinity at depth 3500 metre
753. s00an1.3.txt World Ocean Atlas 2001 1 degree annual analyzed mean - salinity at depth 20 metre
754. s00an1.30.txt World Ocean Atlas 2001 1 degree annual analyzed mean - salinity at depth 4000 metre
755. s00an1.31.txt World Ocean Atlas 2001 1 degree annual analyzed mean - salinity at depth 4500 metre
756. s00an1.32.txt World Ocean Atlas 2001 1 degree annual analyzed mean - salinity at depth 5000 metre
757. s00an1.33.txt World Ocean Atlas 2001 1 degree annual analyzed mean - salinity at depth 5500 metre
758. s00an1.4.txt World Ocean Atlas 2001 1 degree annual analyzed mean - salinity at depth 30 metre
759. s00an1.5.txt World Ocean Atlas 2001 1 degree annual analyzed mean - salinity at depth 50 metre
760. s00an1.6.txt World Ocean Atlas 2001 1 degree annual analyzed mean - salinity at depth 75 metre
761. s00an1.7.txt World Ocean Atlas 2001 1 degree annual analyzed mean - salinity at depth 100 metre
762. s00an1.8.txt World Ocean Atlas 2001 1 degree annual analyzed mean - salinity at depth 125 metre
763. s00an1.9.txt World Ocean Atlas 2001 1 degree annual analyzed mean - salinity at depth 150 metre
764. s00sd1.1.txt World Ocean Atlas 2001 1 degree standard deviation - salinity at depth 0 metre
765. s00sd1.10.txt World Ocean Atlas 2001 1 degree standard deviation - salinity at depth 200 metre
766. s00sd1.11.txt World Ocean Atlas 2001 1 degree standard deviation - salinity at depth 250 metre
767. s00sd1.12.txt World Ocean Atlas 2001 1 degree standard deviation - salinity at depth 300 metre
768. s00sd1.13.txt World Ocean Atlas 2001 1 degree standard deviation - salinity at depth 400 metre
769. s00sd1.14.txt World Ocean Atlas 2001 1 degree standard deviation - salinity at depth 500 metre
770. s00sd1.15.txt World Ocean Atlas 2001 1 degree standard deviation - salinity at depth 600 metre
771. s00sd1.16.txt World Ocean Atlas 2001 1 degree standard deviation - salinity at depth 700 metre

772. s00sd1.17.txt World Ocean Atlas 2001 1 degree standard deviation - salinity at depth 800 metre
773. s00sd1.18.txt World Ocean Atlas 2001 1 degree standard deviation - salinity at depth 900 metre
774. s00sd1.19.txt World Ocean Atlas 2001 1 degree standard deviation - salinity at depth 1000 metre
775. s00sd1.2.txt World Ocean Atlas 2001 1 degree standard deviation - salinity at depth 10 metre
776. s00sd1.20.txt World Ocean Atlas 2001 1 degree standard deviation - salinity at depth 1100 metre
777. s00sd1.21.txt World Ocean Atlas 2001 1 degree standard deviation - salinity at depth 1200 metre
778. s00sd1.22.txt World Ocean Atlas 2001 1 degree standard deviation - salinity at depth 1300 metre
779. s00sd1.23.txt World Ocean Atlas 2001 1 degree standard deviation - salinity at depth 1400 metre
780. s00sd1.24.txt World Ocean Atlas 2001 1 degree standard deviation - salinity at depth 1500 metre
781. s00sd1.25.txt World Ocean Atlas 2001 1 degree standard deviation - salinity at depth 1750 metre
782. s00sd1.26.txt World Ocean Atlas 2001 1 degree standard deviation - salinity at depth 2000 metre
783. s00sd1.27.txt World Ocean Atlas 2001 1 degree standard deviation - salinity at depth 2500 metre
784. s00sd1.28.txt World Ocean Atlas 2001 1 degree standard deviation - salinity at depth 3000 metre
785. s00sd1.29.txt World Ocean Atlas 2001 1 degree standard deviation - salinity at depth 3500 metre
786. s00sd1.3.txt World Ocean Atlas 2001 1 degree standard deviation - salinity at depth 20 metre
787. s00sd1.30.txt World Ocean Atlas 2001 1 degree standard deviation - salinity at depth 4000 metre
788. s00sd1.31.txt World Ocean Atlas 2001 1 degree standard deviation - salinity at depth 4500 metre
789. s00sd1.32.txt World Ocean Atlas 2001 1 degree standard deviation - salinity at depth 5000 metre
790. s00sd1.33.txt World Ocean Atlas 2001 1 degree standard deviation - salinity at depth 5500 metre
791. s00sd1.4.txt World Ocean Atlas 2001 1 degree standard deviation - salinity at depth 30 metre
792. s00sd1.5.txt World Ocean Atlas 2001 1 degree standard deviation - salinity at depth 50 metre
793. s00sd1.6.txt World Ocean Atlas 2001 1 degree standard deviation - salinity at depth 75 metre
794. s00sd1.7.txt World Ocean Atlas 2001 1 degree standard deviation - salinity at depth 100 metre
795. s00sd1.8.txt World Ocean Atlas 2001 1 degree standard deviation - salinity at depth 125 metre
796. s00sd1.9.txt World Ocean Atlas 2001 1 degree standard deviation - salinity at depth 150 metre
797. SALINITY_ANN_AVG.txt Salinity, ann mean
798. SALINITY_GRADIENT.txt Salinity, coastal gradient
799. SALINITY_MAX_MONTH.txt Salinity, max month
800. SALINITY_MIN_MONTH.txt Salinity, min month
801. srzarea.txt Staub and Rosenzweig Zobler Area Codes
802. srzcode.txt Staub and Rosenzweig Zobler Special Codes
803. srzphas.txt Staub and Rosenzweig Zobler Soil Phase Codes
804. srzslop.txt Staub and Rosenzweig Zobler Soil Unit Surface Slope
805. srzsoil.txt Staub and Rosenzweig Zobler Soil Units
806. srzsubs.txt Staub and Rosenzweig Zobler Associated and Included Subsidiary Soil Units
807. srztext.txt Staub and Rosenzweig Zobler Near-Surface Soil Texture
808. SST_INTER_ANNUAL_STDEV.txt SSTemp Inter Annual StDev
809. SST_INTRA_ANN_STDEV.txt SSTemp Intra Annual StDev
810. SST_MAX_MONTH.txt SSTemp, 18yr max monthly
811. SST_MEAN_MONTHLY.txt SSTemp, 18yr mean monthly
812. SST_MEDIAN_MONTHLY.txt SSTemp, 18yr median monthly
813. SST_MIN_MAX_RANGE.txt SSTemp, 18yr min-max range monthly
814. SST_MIN_MONTH.txt SSTemp, 18yr min monthly
815. t00an1.1.txt World Ocean Atlas 2001 1 degree annual analyzed mean - temperature at depth 0 metre
816. t00an1.10.txt World Ocean Atlas 2001 1 degree annual analyzed mean - temperature at depth 200 metre
817. t00an1.11.txt World Ocean Atlas 2001 1 degree annual analyzed mean - temperature at depth 250 metre
818. t00an1.12.txt World Ocean Atlas 2001 1 degree annual analyzed mean - temperature at depth 300 metre
819. t00an1.13.txt World Ocean Atlas 2001 1 degree annual analyzed mean - temperature at depth 400 metre
820. t00an1.14.txt World Ocean Atlas 2001 1 degree annual analyzed mean - temperature at depth 500 metre
821. t00an1.15.txt World Ocean Atlas 2001 1 degree annual analyzed mean - temperature at depth 600 metre
822. t00an1.16.txt World Ocean Atlas 2001 1 degree annual analyzed mean - temperature at depth 700 metre
823. t00an1.17.txt World Ocean Atlas 2001 1 degree annual analyzed mean - temperature at depth 800 metre
824. t00an1.18.txt World Ocean Atlas 2001 1 degree annual analyzed mean - temperature at depth 900 metre
825. t00an1.19.txt World Ocean Atlas 2001 1 degree annual analyzed mean - temperature at depth 1000 metre

826. t00an1.2.txt World Ocean Atlas 2001 1 degree annual analyzed mean - temperature at depth 10 metre
827. t00an1.20.txt World Ocean Atlas 2001 1 degree annual analyzed mean - temperature at depth 1100 metre
828. t00an1.21.txt World Ocean Atlas 2001 1 degree annual analyzed mean - temperature at depth 1200 metre
829. t00an1.22.txt World Ocean Atlas 2001 1 degree annual analyzed mean - temperature at depth 1300 metre
830. t00an1.23.txt World Ocean Atlas 2001 1 degree annual analyzed mean - temperature at depth 1400 metre
831. t00an1.24.txt World Ocean Atlas 2001 1 degree annual analyzed mean - temperature at depth 1500 metre
832. t00an1.25.txt World Ocean Atlas 2001 1 degree annual analyzed mean - temperature at depth 1750 metre
833. t00an1.26.txt World Ocean Atlas 2001 1 degree annual analyzed mean - temperature at depth 2000 metre
834. t00an1.27.txt World Ocean Atlas 2001 1 degree annual analyzed mean - temperature at depth 2500 metre
835. t00an1.28.txt World Ocean Atlas 2001 1 degree annual analyzed mean - temperature at depth 3000 metre
836. t00an1.29.txt World Ocean Atlas 2001 1 degree annual analyzed mean - temperature at depth 3500 metre
837. t00an1.3.txt World Ocean Atlas 2001 1 degree annual analyzed mean - temperature at depth 20 metre
838. t00an1.30.txt World Ocean Atlas 2001 1 degree annual analyzed mean - temperature at depth 4000 metre
839. t00an1.31.txt World Ocean Atlas 2001 1 degree annual analyzed mean - temperature at depth 4500 metre
840. t00an1.32.txt World Ocean Atlas 2001 1 degree annual analyzed mean - temperature at depth 5000 metre
841. t00an1.33.txt World Ocean Atlas 2001 1 degree annual analyzed mean - temperature at depth 5500 metre
842. t00an1.4.txt World Ocean Atlas 2001 1 degree annual analyzed mean - temperature at depth 30 metre
843. t00an1.5.txt World Ocean Atlas 2001 1 degree annual analyzed mean - temperature at depth 50 metre
844. t00an1.6.txt World Ocean Atlas 2001 1 degree annual analyzed mean - temperature at depth 75 metre
845. t00an1.7.txt World Ocean Atlas 2001 1 degree annual analyzed mean - temperature at depth 100 metre
846. t00an1.8.txt World Ocean Atlas 2001 1 degree annual analyzed mean - temperature at depth 125 metre
847. t00an1.9.txt World Ocean Atlas 2001 1 degree annual analyzed mean - temperature at depth 150 metre
848. t00sd1.1.txt World Ocean Atlas 2001 1 degree standard deviation - temperature at depth 0 metre
849. t00sd1.10.txt World Ocean Atlas 2001 1 degree standard deviation - temperature at depth 200 metre
850. t00sd1.11.txt World Ocean Atlas 2001 1 degree standard deviation - temperature at depth 250 metre
851. t00sd1.12.txt World Ocean Atlas 2001 1 degree standard deviation - temperature at depth 300 metre
852. t00sd1.13.txt World Ocean Atlas 2001 1 degree standard deviation - temperature at depth 400 metre
853. t00sd1.14.txt World Ocean Atlas 2001 1 degree standard deviation - temperature at depth 500 metre
854. t00sd1.15.txt World Ocean Atlas 2001 1 degree standard deviation - temperature at depth 600 metre
855. t00sd1.16.txt World Ocean Atlas 2001 1 degree standard deviation - temperature at depth 700 metre
856. t00sd1.17.txt World Ocean Atlas 2001 1 degree standard deviation - temperature at depth 800 metre
857. t00sd1.18.txt World Ocean Atlas 2001 1 degree standard deviation - temperature at depth 900 metre
858. t00sd1.19.txt World Ocean Atlas 2001 1 degree standard deviation - temperature at depth 1000 metre
859. t00sd1.2.txt World Ocean Atlas 2001 1 degree standard deviation - temperature at depth 10 metre
860. t00sd1.20.txt World Ocean Atlas 2001 1 degree standard deviation - temperature at depth 1100 metre
861. t00sd1.21.txt World Ocean Atlas 2001 1 degree standard deviation - temperature at depth 1200 metre
862. t00sd1.22.txt World Ocean Atlas 2001 1 degree standard deviation - temperature at depth 1300 metre
863. t00sd1.23.txt World Ocean Atlas 2001 1 degree standard deviation - temperature at depth 1400 metre
864. t00sd1.24.txt World Ocean Atlas 2001 1 degree standard deviation - temperature at depth 1500 metre
865. t00sd1.25.txt World Ocean Atlas 2001 1 degree standard deviation - temperature at depth 1750 metre
866. t00sd1.26.txt World Ocean Atlas 2001 1 degree standard deviation - temperature at depth 2000 metre
867. t00sd1.27.txt World Ocean Atlas 2001 1 degree standard deviation - temperature at depth 2500 metre
868. t00sd1.28.txt World Ocean Atlas 2001 1 degree standard deviation - temperature at depth 3000 metre
869. t00sd1.29.txt World Ocean Atlas 2001 1 degree standard deviation - temperature at depth 3500 metre
870. t00sd1.3.txt World Ocean Atlas 2001 1 degree standard deviation - temperature at depth 20 metre
871. t00sd1.30.txt World Ocean Atlas 2001 1 degree standard deviation - temperature at depth 4000 metre
872. t00sd1.31.txt World Ocean Atlas 2001 1 degree standard deviation - temperature at depth 4500 metre
873. t00sd1.32.txt World Ocean Atlas 2001 1 degree standard deviation - temperature at depth 5000 metre
874. t00sd1.33.txt World Ocean Atlas 2001 1 degree standard deviation - temperature at depth 5500 metre
875. t00sd1.4.txt World Ocean Atlas 2001 1 degree standard deviation - temperature at depth 30 metre
876. t00sd1.5.txt World Ocean Atlas 2001 1 degree standard deviation - temperature at depth 50 metre
877. t00sd1.6.txt World Ocean Atlas 2001 1 degree standard deviation - temperature at depth 75 metre
878. t00sd1.7.txt World Ocean Atlas 2001 1 degree standard deviation - temperature at depth 100 metre
879. t00sd1.8.txt World Ocean Atlas 2001 1 degree standard deviation - temperature at depth 125 metre

880. t00sd1.9.txt World Ocean Atlas 2001 1 degree standard deviation - temperature at depth 150 metre
881. TIDAL_RANGE.txt Tidal Range
882. tmax_1.txt January Maximum Temperature
883. tmax_10.txt October Maximum Temperature
884. tmax_10_2.5m.txt tmax_10 2.5 minutes
885. tmax_11.txt November Maximum Temperature
886. tmax_11_2.5m.txt tmax_11 2.5 minutes
887. tmax_12.txt December Maximum Temperature
888. tmax_12_2.5m.txt tmax_12 2.5 minutes
889. tmax_1_2.5m.txt tmax_1 2.5 minutes
890. tmax_2.txt February Maximum Temperature
891. tmax_2_2.5m.txt tmax_2 2.5 minutes
892. tmax_3.txt March Maximum Temperature
893. tmax_3_2.5m.txt tmax_3 2.5 minutes
894. tmax_4.txt April Maximum Temperature
895. tmax_4_2.5m.txt tmax_4 2.5 minutes
896. tmax_5.txt May Maximum Temperature
897. tmax_5_2.5m.txt tmax_5 2.5 minutes
898. tmax_6.txt June Maximum Temperature
899. tmax_6_2.5m.txt tmax_6 2.5 minutes
900. tmax_7.txt July Maximum Temperature
901. tmax_7_2.5m.txt tmax_7 2.5 minutes
902. tmax_8.txt August Maximum Temperature
903. tmax_8_2.5m.txt tmax_8 2.5 minutes
904. tmax_9.txt September Maximum Temperature
905. tmax_9_2.5m.txt tmax_9 2.5 minutes
906. tmean_1.txt January Mean Temperature
907. tmean_10.txt October Mean Temperature
908. tmean_10_2.5m.txt tmean_10 2.5 minutes
909. tmean_11.txt November Mean Temperature
910. tmean_11_2.5m.txt tmean_11 2.5 minutes
911. tmean_12.txt December Mean Temperature
912. tmean_12_2.5m.txt tmean_12 2.5 minutes
913. tmean_1_2.5m.txt tmean_1 2.5 minutes
914. tmean_2.txt February Mean Temperature
915. tmean_2_2.5m.txt tmean_2 2.5 minutes
916. tmean_3.txt March Mean Temperature
917. tmean_3_2.5m.txt tmean_3 2.5 minutes
918. tmean_4.txt April Mean Temperature
919. tmean_4_2.5m.txt tmean_4 2.5 minutes
920. tmean_5.txt May Mean Temperature
921. tmean_5_2.5m.txt tmean_5 2.5 minutes
922. tmean_6.txt June Mean Temperature
923. tmean_6_2.5m.txt tmean_6 2.5 minutes
924. tmean_7.txt July Mean Temperature
925. tmean_7_2.5m.txt tmean_7 2.5 minutes
926. tmean_8.txt August Mean Temperature
927. tmean_8_2.5m.txt tmean_8 2.5 minutes
928. tmean_9.txt September Mean Temperature
929. tmean_9_2.5m.txt tmean_9 2.5 minutes
930. tmin_1.txt January Minimum Temperature
931. tmin_10.txt October Minimum Temperature
932. tmin_10_2.5m.txt tmin_10 2.5 minutes
933. tmin_11.txt November Minimum Temperature

934. tmin_11_2.5m.txt tmin_11 2.5 minutes

935. tmin_12.txt December Minimum Temperature

936. tmin_12_2.5m.txt tmin_12 2.5 minutes

937. tmin_1_2.5m.txt tmin_1 2.5 minutes

938. tmin_2.txt February Minimum Temperature

939. tmin_2_2.5m.txt tmin_2 2.5 minutes

940. tmin_3.txt March Minimum Temperature

941. tmin_3_2.5m.txt tmin_3 2.5 minutes

942. tmin_4.txt April Minimum Temperature

943. tmin_4_2.5m.txt tmin_4 2.5 minutes

944. tmin_5.txt May Minimum Temperature

945. tmin_5_2.5m.txt tmin_5 2.5 minutes

946. tmin_6.txt June Minimum Temperature

947. tmin_6_2.5m.txt tmin_6 2.5 minutes

948. tmin_7.txt July Minimum Temperature

949. tmin_7_2.5m.txt tmin_7 2.5 minutes

950. tmin_8.txt August Minimum Temperature

951. tmin_8_2.5m.txt tmin_8 2.5 minutes

952. tmin_9.txt September Minimum Temperature

953. tmin_9_2.5m.txt tmin_9 2.5 minutes

954. treecover.txt Continuous field data - treecover

955. Treecover0.01.txt

956. WAVE_HEIGHT.txt Wave Height

957. whcov1.txt Wilson & Henderson-Sellers Primary Land Cover Classes

958. whcov2.txt Wilson & Henderson-Sellers Secondary Land Cover Classes

959. whlrel.txt Wilson & Henderson-Sellers Land Cover Reliability

960. whsoil.txt Wilson & Henderson-Sellers Code and Properties of Soil Classes

961. whsrel.txt Wilson & Henderson-Sellers Soil Class Reliability

962. wrcla01.txt Webb et al Soil Properties: clay in horizon 1

963. wrcla02.txt Webb et al Soil Properties: clay in horizon 2

964. wrcla03.txt Webb et al Soil Properties: clay in horizon 3

965. wrcla04.txt Webb et al Soil Properties: clay in horizon 4

966. wrcla05.txt Webb et al Soil Properties: clay in horizon 5

967. wrcla06.txt Webb et al Soil Properties: clay in horizon 6

968. wrcla07.txt Webb et al Soil Properties: clay in horizon 7

969. wrcla08.txt Webb et al Soil Properties: clay in horizon 8

970. wrcla09.txt Webb et al Soil Properties: clay in horizon 9

971. wrcla10.txt Webb et al Soil Properties: clay in horizon 10

972. wrcla11.txt Webb et al Soil Properties: clay in horizon 11

973. wrcla12.txt Webb et al Soil Properties: clay in horizon 12

974. wrcla13.txt Webb et al Soil Properties: clay in horizon 13

975. wrcla14.txt Webb et al Soil Properties: clay in horizon 14

976. wrcla15.txt Webb et al Soil Properties: clay in horizon 15

977. wrcont.txt Webb et al Continent Codes from the FAO/UNESCO Soil Map of the World

978. wrdep01.txt Webb et al Soil Properties: depth for horizon 1

979. wrdep02.txt Webb et al Soil Properties: depth for horizon 2

980. wrdep03.txt Webb et al Soil Properties: depth for horizon 3

981. wrdep04.txt Webb et al Soil Properties: depth for horizon 4

982. wrdep05.txt Webb et al Soil Properties: depth for horizon 5

983. wrdep06.txt Webb et al Soil Properties: depth for horizon 6

984. wrdep07.txt Webb et al Soil Properties: depth for horizon 7

985. wrdep08.txt Webb et al Soil Properties: depth for horizon 8

986. wrdep09.txt Webb et al Soil Properties: depth for horizon 9

987. wrdep10.txt Webb et al Soil Properties: depth for horizon 10

988.   wrdep11.txt Webb et al Soil Properties: depth for horizon 11

989.   wrdep12.txt Webb et al Soil Properties: depth for horizon 12

990.   wrdep13.txt Webb et al Soil Properties: depth for horizon 13

991.   wrdep14.txt Webb et al Soil Properties: depth for horizon 14

992.   wrdep15.txt Webb et al Soil Properties: depth for horizon 15

993.   wrmodii.txt Webb et al Model II Soil Water (mm)

994.   wrprof.txt Webb et al Potential Storage of Water in Soil Profile (mm)

995.   wrroot.txt Webb et al Potential Storage of Water in Root Zone (mm)

996.   wrsan01.txt Webb et al Soil Properties: sand in horizon 1

997.   wrsan02.txt Webb et al Soil Properties: sand in horizon 2

998.   wrsan03.txt Webb et al Soil Properties: sand in horizon 3

999.   wrsan04.txt Webb et al Soil Properties: sand in horizon 4

1000. wrsan05.txt Webb et al Soil Properties: sand in horizon 5

1001. wrsan06.txt Webb et al Soil Properties: sand in horizon 6

1002. wrsan07.txt Webb et al Soil Properties: sand in horizon 7

1003. wrsan08.txt Webb et al Soil Properties: sand in horizon 8

1004. wrsan09.txt Webb et al Soil Properties: sand in horizon 9

1005. wrsan10.txt Webb et al Soil Properties: sand in horizon 10

1006. wrsan11.txt Webb et al Soil Properties: sand in horizon 11

1007. wrsan12.txt Webb et al Soil Properties: sand in horizon 12

1008. wrsan13.txt Webb et al Soil Properties: sand in horizon 13

1009. wrsan14.txt Webb et al Soil Properties: sand in horizon 14

1010. wrsan15.txt Webb et al Soil Properties: sand in horizon 15

1011. wrsil01.txt Webb et al Soil Properties: silt in horizon 1

1012. wrsil02.txt Webb et al Soil Properties: silt in horizon 2

1013. wrsil03.txt Webb et al Soil Properties: silt in horizon 3

1014. wrsil04.txt Webb et al Soil Properties: silt in horizon 4

1015. wrsil05.txt Webb et al Soil Properties: silt in horizon 5

1016. wrsil06.txt Webb et al Soil Properties: silt in horizon 6

1017. wrsil07.txt Webb et al Soil Properties: silt in horizon 7

1018. wrsil08.txt Webb et al Soil Properties: silt in horizon 8

1019. wrsil09.txt Webb et al Soil Properties: silt in horizon 9

1020. wrsil10.txt Webb et al Soil Properties: silt in horizon 10

1021. wrsil11.txt Webb et al Soil Properties: silt in horizon 11

1022. wrsil12.txt Webb et al Soil Properties: silt in horizon 12

1023. wrsil13.txt Webb et al Soil Properties: silt in horizon 13

1024. wrsil14.txt Webb et al Soil Properties: silt in horizon 14

1025. wrsil15.txt Webb et al Soil Properties: silt in horizon 15

1026. wrsoil.txt Webb et al Soil Profile Thickness (cm)

1027. wrtext.txt Webb et al Texture-Based Potential Storage of Water (mm)

1028. wrzsoil.txt Webb et al Soil Particle Size Properties Zobler Soil Types

1029. x00an1.1.txt World Ocean Atlas 2001 1 degree annual analyzed mean - percent oxygen saturation at
       depth 0 metre

1030. x00an1.10.txt World Ocean Atlas 2001 1 degree annual analyzed mean - percent oxygen saturation at
       depth 200 metre

1031. x00an1.11.txt World Ocean Atlas 2001 1 degree annual analyzed mean - percent oxygen saturation at
       depth 250 metre

1032. x00an1.12.txt World Ocean Atlas 2001 1 degree annual analyzed mean - percent oxygen saturation at
       depth 300 metre

1033. x00an1.13.txt World Ocean Atlas 2001 1 degree annual analyzed mean - percent oxygen saturation at
       depth 400 metre

1034. x00an1.14.txt World Ocean Atlas 2001 1 degree annual analyzed mean - percent oxygen saturation at
       depth 500 metre

1035. x00an1.15.txt World Ocean Atlas 2001 1 degree annual analyzed mean - percent oxygen saturation at
       depth 600 metre

1036. x00an1.16.txt World Ocean Atlas 2001 1 degree annual analyzed mean - percent oxygen saturation at
       depth 700 metre

1037. x00an1.17.txt World Ocean Atlas 2001 1 degree annual analyzed mean - percent oxygen saturation at depth 800 metre

1038. x00an1.18.txt World Ocean Atlas 2001 1 degree annual analyzed mean - percent oxygen saturation at depth 900 metre

1039. x00an1.19.txt World Ocean Atlas 2001 1 degree annual analyzed mean - percent oxygen saturation at depth 1000 metre

1040. x00an1.2.txt World Ocean Atlas 2001 1 degree annual analyzed mean - percent oxygen saturation at depth 10 metre

1041. x00an1.20.txt World Ocean Atlas 2001 1 degree annual analyzed mean - percent oxygen saturation at depth 1100 metre

1042. x00an1.21.txt World Ocean Atlas 2001 1 degree annual analyzed mean - percent oxygen saturation at depth 1200 metre

1043. x00an1.22.txt World Ocean Atlas 2001 1 degree annual analyzed mean - percent oxygen saturation at depth 1300 metre

1044. x00an1.23.txt World Ocean Atlas 2001 1 degree annual analyzed mean - percent oxygen saturation at depth 1400 metre

1045. x00an1.24.txt World Ocean Atlas 2001 1 degree annual analyzed mean - percent oxygen saturation at depth 1500 metre

1046. x00an1.25.txt World Ocean Atlas 2001 1 degree annual analyzed mean - percent oxygen saturation at depth 1750 metre

1047. x00an1.26.txt World Ocean Atlas 2001 1 degree annual analyzed mean - percent oxygen saturation at depth 2000 metre

1048. x00an1.27.txt World Ocean Atlas 2001 1 degree annual analyzed mean - percent oxygen saturation at depth 2500 metre

1049. x00an1.28.txt World Ocean Atlas 2001 1 degree annual analyzed mean - percent oxygen saturation at depth 3000 metre

1050. x00an1.29.txt World Ocean Atlas 2001 1 degree annual analyzed mean - percent oxygen saturation at depth 3500 metre

1051. x00an1.3.txt World Ocean Atlas 2001 1 degree annual analyzed mean - percent oxygen saturation at depth 20 metre

1052. x00an1.30.txt World Ocean Atlas 2001 1 degree annual analyzed mean - percent oxygen saturation at depth 4000 metre

1053. x00an1.31.txt World Ocean Atlas 2001 1 degree annual analyzed mean - percent oxygen saturation at depth 4500 metre

1054. x00an1.32.txt World Ocean Atlas 2001 1 degree annual analyzed mean - percent oxygen saturation at depth 5000 metre

1055. x00an1.33.txt World Ocean Atlas 2001 1 degree annual analyzed mean - percent oxygen saturation at depth 5500 metre

1056. x00an1.4.txt World Ocean Atlas 2001 1 degree annual analyzed mean - percent oxygen saturation at depth 30 metre

1057. x00an1.5.txt World Ocean Atlas 2001 1 degree annual analyzed mean - percent oxygen saturation at depth 50 metre

1058. x00an1.6.txt World Ocean Atlas 2001 1 degree annual analyzed mean - percent oxygen saturation at depth 75 metre

1059. x00an1.7.txt World Ocean Atlas 2001 1 degree annual analyzed mean - percent oxygen saturation at depth 100 metre

1060. x00an1.8.txt World Ocean Atlas 2001 1 degree annual analyzed mean - percent oxygen saturation at depth 125 metre

1061. x00an1.9.txt World Ocean Atlas 2001 1 degree annual analyzed mean - percent oxygen saturation at depth 150 metre

1062. x00sd1.1.txt World Ocean Atlas 2001 1 degree standard deviation - percent oxygen saturation at depth 0 metre

1063. x00sd1.10.txt World Ocean Atlas 2001 1 degree standard deviation - percent oxygen saturation at depth 200 metre

1064. x00sd1.11.txt World Ocean Atlas 2001 1 degree standard deviation - percent oxygen saturation at depth 250 metre

1065. x00sd1.12.txt World Ocean Atlas 2001 1 degree standard deviation - percent oxygen saturation at depth 300 metre

1066. x00sd1.13.txt World Ocean Atlas 2001 1 degree standard deviation - percent oxygen saturation at depth 400 metre

1067. x00sd1.14.txt World Ocean Atlas 2001 1 degree standard deviation - percent oxygen saturation at depth 500 metre

1068. x00sd1.15.txt World Ocean Atlas 2001 1 degree standard deviation - percent oxygen saturation at depth 600 metre

1069. x00sd1.16.txt World Ocean Atlas 2001 1 degree standard deviation - percent oxygen saturation at depth 700 metre

1070. x00sd1.17.txt World Ocean Atlas 2001 1 degree standard deviation - percent oxygen saturation at depth 800 metre

1071. x00sd1.18.txt World Ocean Atlas 2001 1 degree standard deviation - percent oxygen saturation at depth 900 metre

1072. x00sd1.19.txt World Ocean Atlas 2001 1 degree standard deviation - percent oxygen saturation at depth 1000 metre

1073. x00sd1.2.txt World Ocean Atlas 2001 1 degree standard deviation - percent oxygen saturation at depth 10 metre

1074. x00sd1.20.txt World Ocean Atlas 2001 1 degree standard deviation - percent oxygen saturation at depth 1100 metre

1075. x00sd1.21.txt World Ocean Atlas 2001 1 degree standard deviation - percent oxygen saturation at depth 1200 metre

1076. x00sd1.22.txt World Ocean Atlas 2001 1 degree standard deviation - percent oxygen saturation at depth 1300 metre

1077. x00sd1.23.txt World Ocean Atlas 2001 1 degree standard deviation - percent oxygen saturation at depth 1400 metre

1078. x00sd1.24.txt World Ocean Atlas 2001 1 degree standard deviation - percent oxygen saturation at depth 1500 metre

1079. x00sd1.25.txt World Ocean Atlas 2001 1 degree standard deviation - percent oxygen saturation at depth 1750 metre

1080. x00sd1.26.txt World Ocean Atlas 2001 1 degree standard deviation - percent oxygen saturation at depth 2000 metre

1081. x00sd1.27.txt World Ocean Atlas 2001 1 degree standard deviation - percent oxygen saturation at depth 2500 metre

1082. x00sd1.28.txt World Ocean Atlas 2001 1 degree standard deviation - percent oxygen saturation at depth 3000 metre

1083. x00sd1.29.txt World Ocean Atlas 2001 1 degree standard deviation - percent oxygen saturation at depth 3500 metre

1084. x00sd1.3.txt World Ocean Atlas 2001 1 degree standard deviation - percent oxygen saturation at depth 20 metre

1085. x00sd1.30.txt World Ocean Atlas 2001 1 degree standard deviation - percent oxygen saturation at depth 4000 metre

1086. x00sd1.31.txt World Ocean Atlas 2001 1 degree standard deviation - percent oxygen saturation at depth 4500 metre

1087. x00sd1.32.txt World Ocean Atlas 2001 1 degree standard deviation - percent oxygen saturation at depth 5000 metre

1088. x00sd1.33.txt World Ocean Atlas 2001 1 degree standard deviation - percent oxygen saturation at depth 5500 metre

1089. x00sd1.4.txt World Ocean Atlas 2001 1 degree standard deviation - percent oxygen saturation at depth 30 metre

1090. x00sd1.5.txt World Ocean Atlas 2001 1 degree standard deviation - percent oxygen saturation at depth 50 metre

1091. x00sd1.6.txt World Ocean Atlas 2001 1 degree standard deviation - percent oxygen saturation at depth 75 metre

1092. x00sd1.7.txt World Ocean Atlas 2001 1 degree standard deviation - percent oxygen saturation at depth 100 metre

1093. x00sd1.8.txt World Ocean Atlas 2001 1 degree standard deviation - percent oxygen saturation at depth 125 metre

1094. x00sd1.9.txt World Ocean Atlas 2001 1 degree standard deviation - percent oxygen saturation at depth 150 metre

5.1.1   Global ecosystems database

The Global Ecosystems Database (GED) project was developed by the National Geophysical Data Center (NGDC) of the U.S. National Oceanic and Atmospheric Administration (NOAA), and the U.S. Environmental Protection Agency's (EPA) Environmental Research Laboratory in Corvallis, Oregon (ERL-C) [KHO+00]. The GED is the most extensive and accessible of global databases for niche modeling with datasets including climate, ecoregions, land use, soils, topography. Some themes such as climate are available in alternative calibrations, and from more than one author.

5.1.2 Worldclim

Worldclim is a set of global climate layers (grids) on a square kilometer grid aimed at representing more biologically meaningful variables [HCP$^+$05]. The bioclimatic variables represent physiologically relevant annual trends (e.g., mean annual temperature, annual precipitation), seasonality (e.g., annual range in temperature and precipitation) and extreme or limiting environmental factors (e.g., temperature of the coldest and warmest month, and precipitation of the wet and dry quarters) as follows:

BIO1 Annual Mean Temperature

BIO2 Mean Diurnal Range (Mean of monthly (max temp - min temp))

BIO3 Isothermality (P2/P7) (* 100)

BIO4 Temperature Seasonality (standard deviation *100)

BIO5 Max Temperature of Warmest Month

BIO6 Min Temperature of Coldest Month

BIO7 Temperature Annual Range (P5-P6)

BIO8 Mean Temperature of Wettest Quarter

BIO9 Mean Temperature of Driest Quarter

BIO10 Mean Temperature of Warmest Quarter

BIO11 Mean Temperature of Coldest Quarter

BIO12 Annual Precipitation

BIO13 Precipitation of Wettest Month

BIO14 Precipitation of Driest Month

BIO15 Precipitation Seasonality (Coefficient of Variation)

BIO16 Precipitation of Wettest Quarter

BIO17 Precipitation of Driest Quarter

BIO18 Precipitation of Warmest Quarter

BIO19 Precipitation of Coldest Quarter

5.1.3   World ocean atlas

The World Ocean Atlas 2001 (WOA01) Data for Ocean Data View are global ocean historical hydrographic data from the U.S. NODC World Ocean Atlas 2001 [CAB+02]. The data are on one and five degree grid squares at the following standard depths (in m): 0, 10, 20, 30, 50, 75, 100, 125, 150, 200, 250, 300, 400, 500, 600, 700, 800, 900, 1000, 1100, 1200, 1300, 1400, 1500, 1750, 2000, 2500, 3000, 3500, 4000, 4500, 5000, 5500. Data are available for the following variables:

- Temperature [C]

- Salinity [psu]

- Oxygen [ml/l]

- Oxygen Saturation [%]

- AOU [ml/l]

- Phosphate [mmol/l]

- Nitrate [mmol/l]

- Silicate [mmol/l]

5.1.4   Continuous fields

The Continuous Fields 1 Km Tree Cover was developed at the University of Maryland representing land cover as continuous fields of vegetation characteristics using a linear mixture model approach [HDT+03]. These datasets provide satellite imagery of vegetation cover in the form of continuous variables of plant types which should be more amenable to model development using linear modeling approaches such as generalized linear models. This approach differs from conventional vegetation maps based on vegetation classification schemes producing categorical variables. The dataset contains 1km cells estimating:

- Percent tree cover

- Percentage cover leaf longevity, evergreen

- Percentage cover for leaf longevity, deciduous

- Percentage cover for leaf type, broadleaf

- Percentage cover for leaf type, needleleaf

5.1.5   Hydro1km

The Hydro1km dataset was developed at the U.S. Geological Survey's Center for Earth Resources Observation and Science (EROS), to provide comprehensive and consistent global coverage of topographically derived data sets, including streams, drainage basins and ancillary layers derived from the USGS' 30 arc-second digital elevation model of the world. The HYDRO1k provides data including:

- Elevation

- Aspect

- Flow Accumulation

- Compound Topographic Index

- Drainage Basins

- Slope

- Flow Direction

- Streams

5.1.6   WhyWhere

A number of datasets are available on the web from free sources. Unfortunately they come in a variety of formats and typically need to be processed into a common format before they can be used. The WhyWhere archive is a compilation of the datasets into a consistent format.

Here we describe the use of the Storage Resource Broker (SRB) to support data intensive approaches to Environmental Niche Modeling (ENM) by providing access to cropped images from a remote SRB data store of almost one thousand global coverage datasets in a standard format. The archive is currently used in, but not restricted to, the data mining algorithm WhyWhere. The basic architecture of the system is illustrated below (Figure 5.1).

5.2   Archives

Future needs are many if data archives are to be used as a production resource for the general research community. The following are some of the main challenges:

5.2.1 Traffic

If traffic is large, each analysis requires download and processing of a large section of the collection by each researcher each time they start analysis. Replication of archives and a means for the client to select the most efficient (e.g. proximate) archive for connection and download helps to achieve this.

5.2.2 Management

If the resource is widely used to support research, attention will be paid to management of the archive by a community of researchers for maintaining, updating, cleaning and improving the archive. Models of funding need to be in place for that to be achieved on a long term basis.

5.2.3 Interaction

The needs and solutions that drive the environmental data archive component are not necessarily the same as those of the spatial mining algorithms. For example, server side cropping operations avoid the overhead of downloading the whole dataset into the client.

5.2.4 Updating

Where datasets are obtained from other sources of data, regular updating schemes provide timeliness while maintaining and documenting past versions in case they are needed.

5.2.5 Legacy

Solutions must be integrated with existing systems. For example, a remote partial file transfer capability to be accomplished by special operations within the server could use the Open GIS Consortium (OGC) geographic specification to provide a 'wrapper' for accessing data that is shared by other archives.

Despite the costs in setting up and operating an archive, these are amortized over time by the many benefits. More advanced systems contain such features as location encapsulation for archived files and metadata, so that programs and users don't need to know where the actual data are stored. Set up as distributed data management system, they can provide virtually unlimited storage space for geospatial data/images. Archives also provide automated database backup, replication, and encryption and security control.

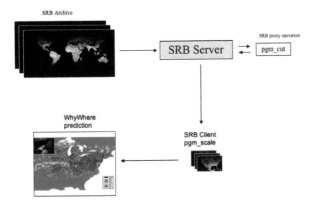

FIGURE 5.1: The components and operation of the WhyWhere SRB data archive for ecological niche modeling.

5.2.6 Example: WhyWhere archive

Figure 5.1 illustrates the components and operation of the WhyWhere SRB data archive for ecological niche modeling. A large set of images and meta data are stored in a central archive. The client directs the server to crop an image in the archive using a server-side proxy operation. The cropped image is copied to the local directory and scaled by the client to the resolution required for the prediction algorithm.

5.2.7 Browsing

While some web-based tools such as ftp can be used to browse collections at single locations, specialized tools are needed for collections, not stored on a single server. There tools use of metadata to construct and maintain a coherent view of a collection. One such tool is *inQ* and can be used to browse the WhyWhere collection using the following settings:

```
Name: testuser
Host: orion.sdsc.edu
Domain: sdsc
Port: 7613
Authorization: ENCRYPT1
Password: TESTUSER
```

5.2.8   Format

The main elements of the data archive are as follows:

- a single pgm image format for at least 1000 data variables,
- all data sets of global extent but variable scale,
- all data described with meta data, and
- capacity to supply variables cropped and scaled to a specific size and resolution.

The pgm image format used to store the variables has strengths and weaknesses. Why categorical variables are limited in number of categories, most categorical variables describing vegetation, landscape or soil types contain fewer than 256 values. Continuous value variables were normalized between 0 and 256 according to their maximum and minimum values. While this trade off for efficiency resulted in loss of information, it is not as problematic as it might seem given the correlative algorithms used in mining the dataset are only looking for statistical associations.

5.2.9   Meta data

The meta data format consists of simple attribute/value pairs as used in the Global Ecosystem Database (GED). An example is shown below. While more complex meta data would be useful, this format was adequate for most purposes. The meta data documents images, and provides the dimensions to allow extraction by the cropping algorithm so that parts of the image file can be accessed.

```
file title : Legates and Willmott Annual Temperature (0.1C)
data type : integer
file type : binary
columns : 720
rows : 360
ref. system : lat/long
ref. units : deg
unit dist. : 1.0000000
min. X : -180.0000000
max. X : 180.0000000
min. Y : -90.0000000
max. Y : 90.0000000
posn error : unknown
resolution : 0.5000000
```

```
min. value : -569
max. value : 299
value units : 0.1 degrees Celsius
value error : unknown
```

5.2.10   Operations

In developing a generic approach to accessing the geographic data sets, one needs to pay attention to increased usage in the future and the options for balancing the client/server load. One approach is to download the whole file into a local machine and operate on the image locally. This is not a good solution in terms of performance as many files are greater than 1GByte and only partial data is needed.

Methods such as grid-ftp provide partial file transfer, where client side operations (grid-ftp client) repeatedly calculate offsets and then make grid-ftp calls with new offsets and numbers of bytes for data transfer. In this case, a customized client assembles the resulting lines of the data to create the dataset.

The problem with this approach is that data transfer is very slow, presumably due to latency of the Internet connections that are made and broken for each line of the image. By developing a cropping function on the server side, and keeping the scaling function on the client side, the SRB functionality is extended appropriately. The server side reads the header of the image file, retrieves and assembles a series of lines from the file corresponding to the area needed then passes it to the calling client. There two operations are exemplified by the image processing library netpbm operations *pgmcut*, producing a partial image, and *pgmscale* either increasing or reducing the size.

We found this approach to give reasonable performance and is the 'generic' solution currently used in the WhyWhere application. The SRB connection is carried out as a proxy operation on the server through the use of a program in the client called *rpgm*.

5.3   Summary

The purpose of this section has been to show that management of data for niche modeling goes beyond simple approaches consisting of a set of files in a local directory. Only substandard results can be expected from relying on a few arbitrarily chosen and unknown quality ad hoc datasets. A wide variety of datasets are currently available. Better quality niche modeling will result from using data acquired from true archives – shared by many studies and trusted with the highest level of quality.

New technologies can assist inter-institutional coordination of larger, better-characterized and maintained, integrated biological data resources with high bandwidth links and high update frequency. Successful niche modeling will require greater proficiency in the use of new middleware tools for handling meta data. These tools should support meta data integration and the functional interaction between databases, tracking changes in the data, data model descriptions, mismatches and types of error, in order to monitor reliability of the resources.

Chapter 6

Examples

In this chapter are some examples of novel applications of niche modeling, selected primarily to illustrate important counterexamples. Most competent niche model studies model the response of the species as an 'inverted U', and not linear responses. These examples illustrate some more unusual responses and uses of response functions.

First, we discuss a statistic for evaluating the skill of a niche model based on the idea of a 'draw' from a population. The examples follow, illustrating the steps followed in developing a niche model: the use of presence only or presence/absence data, the use of a mask, and the selection of variables. There are no hard and fast rules about these alternatives, but they do influence results.

The examples are:

- house price increase,
- Brown Treesnake threat, and
- zebra mussel spread.

Obviously house price increases is not a biological subject but is included to illustrate the generality of the approach.

6.0.1 Model skill

Accuracy is the observed proportion of predicted values correct. Perfect accuracy would look like Table 6.1.

However, the effective accuracy is only relative to the result expected if those predictions were made at random. For example, if 100% of test items were labeled P then achieving perfect accuracy would require no skill.

There are many way of determining the skill of models. In predicting presence and absence, statistics such as the Receiver Operating Statistic (ROC) [HM82] are effective and popular. Considering the specific case of randomly selected presence and absence points, the area under the ROC curve is a measure of the probability that the model correctly identifies P and A.

TABLE 6.1:
Example of a
contingency
table with a
perfect score.

	P	A
P	1.00	0.00
A	0.00	1.00

In presence/absence analysis (PA) the absence points are explicitly specified, but often absence points are not available. In presence-only analysis (P), the absence points are generated in some way, typically by sampling all points ('background') at random. When only presence points are available, P analysis must be used with such statistics as the ROC.

A review of the ways of quantifying the skill of a model is not attempted here, as an alternate way of viewing the problem is presented. In this view, the survey of the species can be seen as a draw from a population, where the population is the distribution of the species in the environment. Seen in this way, appropriate measures for quantifying the skill of models compare the distributions between what is observed and what would be expected if the draw were random.

For example, consider an environmental variable not related to the distribution of a species. The distribution of environmental values in those points where the species occurs should be identical to the distribution of environmental values in the whole region. But if the observed distribution of values differs significantly from the expected distribution, there is a basis for suspecting the occurrence of the species is related to that variable.

There are three possibilities:

- insignificant variables

- significant variables

- most significant variables

Providing the set of observations of the species is large enough, and the set of variables includes those potentially causing the observed patterns in the species, it may be inferred that the variables with the most significant differences are causing the distribution. These will also be most skillful at estimating the probability of finding the species in new areas.

Thus in this view, points of presence and points of absence are not complements as is usually assumed. Here, absence points are viewed as a separate draw from a population, and potentially identifiable as a separate entity.

6.0.2 Calculating accuracy

Where categories of environmental values are coded as colors in an image, the sets of presence points p, or background points b for determining model skill have probabilities as follows.

B_i is the proportion of b in the background of a specific color i.

P_i is the proportion of each occurrence point p of a color i.

The Chi Square test for differences between distribution applied over environmental categories is one way to quantify model skill:

$$\chi^2 = \sum_{i=1}^{n} \frac{(pP_i - bB_i)^2}{bB_i}$$

The expected accuracy Pr is a simple and effective indicator of skill. To calculate Pr sum the maximum of the proportions of presence or absence proportions over each environmental category, and divide by two.

$$Pr = \sum_{i=1}^{n} \frac{max(P_i, B_i)}{2}$$

This analog of accuracy is derived from the expected accuracy of a decision rule where a species is predicted present if in category i if $P_i > B_i$, and absence otherwise.

Pr has an expected value of 0.5 when there is no relationship between the variable and the draw, and approaches the maximum value of one when all the drawn points fall into a single category in a large number of environmental categories.

6.1 Predicting house prices

Predicting prices of real estate is similar to predicting species distributions. Niche models can be developed if locations such as cities and their house prices or increases are correlated with environmental variables. Here we model and predict the increase in house prices in metro areas of the United States in 2004.

The National Association of Realtors publishes housing statistics for metropolitan areas, as an excel spreadsheet of percent change in the median price of houses for past years and the last four past quarters.

```
Metropolitan_Area, 2002, 2003, 2004, 2004:III, 2004:IV, 2005:I,
2005:II, 2005:III "Allentown-Bethlehem-Easton, PA-NJ", 161.1,
184.7, 207.3, 222.6, 210.5, 214.8, 242.7, N/A, N/A
...
```

To perform the analysis we need the decimal coordinates of each metropoliton area. For example:

```
-117.425 47.65888889
-74.42333333 39.36416667
```

A free server is available at http://geocoder.us for obtaining these coordinates. This server returns the latitude & longitude of any US address. Here is an example of a query for Phoenix, AZ where price of houses increased 55% in 2004:

```
POST: http://geocoder.us/service/csv/geocode?city=Phoenix
&state=AZ
RETURNS: -112.0733333,33.44833333,Maricopa, Phoenix, AZ
```

The next step is to extract coordinates of metro areas showing a median price increase greater than the value of interest. In this example, there are 24 metro areas with median house increases greater than 20%. These points are pasted into the prediction application. All the data were used in the prediction, and none were held back for validation. With this small number of points, there would have been substantial variation between variables selected due to subsetting of points. Had I been interested in more rigorous tests of statistical skill I would have repeated the analysis a number of times, estimating accuracy on a 'held back' set of points.

6.1.1   Analysis

The following are the results of predicting the distribution of metro areas with increases greater than 20% in 2004. Some experiments show the consequence of these choices later.

- dataset size - all terrestrial vs only climate variables

- data type - annual climate vs monthly and other climate

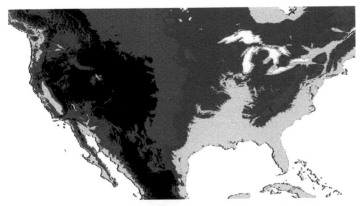

FIGURE 6.1: Predicted price increases >20% using altitude 2.5 minute variable selected by WhyWhere from the dataset of 528 All Terrestrial variables.

- mask - the areas of ocean are not included in the analysis
- P analysis - the distribution of the P draw is compared with the B population.

These were done using WhyWhere with the All Terrestrial dataset consisting of 528 terrestrial variables the analysis used defaults associated with this software. The results were as follows:

```
Environmental Data from All Terrestrial (528)
alt_2.5m: altitude 2.5 minutes resolution
Accuracy 0.805
```

In the predicted distribution shown on Figure 6.1 the lighter the area the higher the predicted probability. At this fine scale the points of the metro areas are hard to see, but will be more apparent on later images.

For comparison here are the results with fewer climate related variables run on a smaller data set of climate variables called ClimateAnnAve consisting of annual temperature, precipitation and standard deviations. The accuracies and maps are as follows:

```
Environmental Data for .0 from ClimateAnnAv
0. lwcpr00 Legates Willmott Annual Corrected Precipitation (mm/year)
Range 0 to 6626 millimeters/year
Accuracy 0.787
```

The results show annual precipitation is the best predictor (Table 6.2). Figure 6.2 shows the predictions resulting from the model as derived from

TABLE 6.2:   Accuracies for each single variable of Legates
Willmott Annual temperatures and precipitation data.

	Variable	Code	Value
1	Corrected Precipitation (mm/year)	lwcpr00.pgm	0.80
2	Corrected Precipitation (std. dev.)	lwcsd00.pgm	0.67
3	Standard Error (mm/year)	lwerr00.pgm	0.74
4	Measured Precipitation (mm/year)	lwmpr00.pgm	0.78
5	Measured Precipitation (std. dev.)	lwmsd00.pgm	0.69
6	Temperature (0.1C)	lwtmp00.pgm	0.75
7	Temperature (std. dev.)	lwtsd00.pgm	0.65

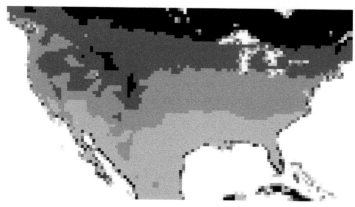

FIGURE 6.2: Predicted price increases greater than 20% using annual
climate averages and presence only data

the data in the histogram of the proportion of grid cells in the precipitation
variable (Figure 6.3). The histogram shows that proportions of precipitation
values in locations where metro areas with appreciation greater than 20% as
a solid line and the proportion of values of precipitation for the entire area as
a dashed line.

   Examination of the histogram shows the response is bimodal in form (Figure 6.3). High price increases occurred in areas of either high summer (eastern
US) or low summer (western US) precipitation. As a further test, we could
predict those 78 metro areas that had an increase of less than 10% per year.
The results were as follows (Figure 6.4):

```
Environmental Data for .3 from ClimateAnnAv
3. lwmpr00 Legates Willmott Annual Measured Precipitation
(mm/year)
Range 0 to 6434 millimeters/year
Accuracy 0.732
```

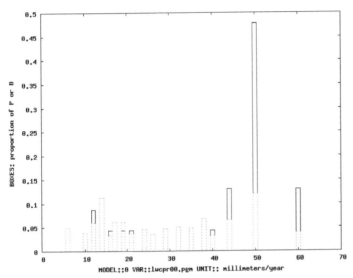

FIGURE 6.3:   Frequency of P and B environmental values for precipitation. The histogram of the proportion of grid cells in the precipitation variable in the locations where metro areas with appreciation greater than 20% (solid line showing presence or P points) and the proportion of values of precipitation for the entire area (dashed line showing background B).

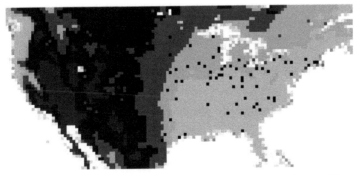

FIGURE 6.4:   Predicted price increases of less than 10% with locations as black squares.

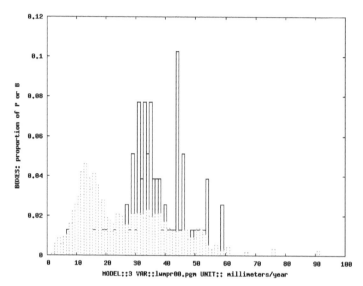

FIGURE 6.5: Frequency of environmental variables predicting house price increases <10%. Note in this case the response if the P point (solid lines) is unimodal.

As expected the precipitation is again most accurate but is no longer bimodal in form. Low growth in prices occurred in areas of moderate precipitation, with high variance in areas of high precipitation. This is the response pattern most often seen when predicting species.

6.1.2   P data and no mask

Here environmental dataset used was called All_Terrestrial and consisted of 205 variables. The mask is the area deliberately excluded from analysis. For example, in analyzing a terrestrial species one might exclude all areas of water, or at least ocean. The converse would be true of marine species. Reducing the area of interest typically allows the model to produce more relevant predictors to the area of interest.

The results were achieved using only presence points and no mask were:

```
Environmental Data for .22.90 from AllTerrestrial
22. lcprc02 Leemans and Cramer February Precipitation (mm/month)
Range 0 to 652 millimeters/month
90. lwmpr06 Legates Willmott June Measured Precipitation
(mm/month)
Range 0 to 1129 millimeters/month
Accuracy 0.921
```

The environmental category for ocean can be seen as the tallest bar in the histogram of environmental categories. The variable selected for precipitation - lcprc02 - is constant across the ocean allowing it to be treated as a single environmental category containing no house price points.

Now we mask out the oceans. This is because, firstly, we are not interested in them. Secondly, if included, the first variable selected often must distinguish between ocean and land. In the previous analysis where the ocean was not masked the model may have selected the first variable primarily in order to discriminate between ocean and land. Given we already know this, the first variable could be regarded as redundant information.

The results with the ocean mask are:

```
Environmental Data for .185.73 from AllTerrestrial
185. alt Altitude
Range to percentage
73. lwcsd02 Legates Willmott February Corrected Precipitation
(std. dev.)
Range 0 to 176 millimeters/month
Accuracy 0.85
```

Masking out ocean results in different variables being selected. Altitude is the first variable selected, possibly because metro areas with high growth have low altitude. Clearly altitude would not have been selected without the mask, as oceans and high appreciation metro areas both have low altitude.

Masking also lowers the accuracy (0.85 vs 0.92). The task of the model is easier when assessment of accuracy includes the absence points produced by large areas of ocean. When the oceans are masked out, accuracy is calculated only on those areas where cities are possible, i.e. on dry land. The task of the model is a little harder; the proportion of points predicted accurately is reduced.

Thus lack of appropriate masking can lead to exaggeration of the accuracy of models.

6.1.3 Presence and absence (PA) data

While the following analysis compared the distribution of environmental value points of high house value appreciation, and found low altitude to be a good predictor, this might be because most metro areas are located at low altitude. PA analysis is primarily concerned with distinguishing between the metro areas of high and low appreciation. To perform PA analysis in WhyWhere, we append a 1 or 0 depending on P or A to all coordinates of metro areas. On re-running the analysis we obtain:

```
Environmental Data for .195.84 from AllTerrestrial
195. bio18 Precipitation of Warmest Quarter
Range to percentage
84. lwmpr00 Legates Willmott Annual Measured Precipitation
(mm/year)
Range 0 to 6434 millimeters/year
Accuracy 0.818
```

6.1.4 Interpretation

Caution needs to be exercised in interpreting house price increases in terms of niche models. The climatic variables used are not the market-based variables typically associated with price increases. On the one hand this could be interpreted as a demonstration that climate variables are not appropriate for all predicted entities. For general niche modeling a larger set, including such variables as altitude, is necessary.

On the other hand, the result could also be interpreted as contradicting the common wisdom that a major real estate trend is 'baby boomers' heading for warmer climates. That is, the commentary that regards the real estate expansion related to temperature, largely due to growth in Florida and the SW, is clearly contradicted by the increase in markets in the NE and NW of the USA. Also, many high temperature regions are not increasing in price.

This second interpretation is only adequate in the context of comparing temperature and rainfall. However, when a larger set of variables is used, an entirely different variable is selected as the most accurate, suggesting an entirely different explanation for growth.

Secondly, the histograms of responses show growth is in two groups, high and low rainfall. That a bimodal variable is identified as the best predictor shows that sometimes a unimodal response curve is not adequate for niche modeling.

If we had tried to fit a uni-modal curve, as is frequently the case with species data and generalized linear modeling, we would have identified a different, less accurate variable predicting house price increases. This example demonstrates that the capacity for more general distribution models is required for niche modeling in more general application areas.

The analysis may have produced a bimodal result either because the response is complex of two factors, i.e. it may be that they have advanced in two different regions for two different reasons, or there may simply be no variable available that models price increases adequately. It could also be thought that the metro areas with intermediate growth in the mid-continent represent the actual stable niche, and the areas with precipitation extremes have more forces creating unstable prices, such as building space limitations or high building material prices.

6.2 Brown Treesnake

The Brown Treesnake (BTS) is a significant threat to many native species where it has invaded. There are a large number of ongoing efforts to prevent its movement into Florida, Hawaii, Texas, and other potential new habitats.

The BTS (family Colubridae) belongs to the genus Boiga, a group of about 25 species that are referred to as 'cat-eyed' snakes due to their vertical pupil. They are rear fanged, have a large head in relation to its body, brownish or greenish, sometimes faint bands. Adults are generally 4 -5 feet long. BTS can survive for extended periods of time without food, a trait that enables it to survive in ship-bound and stored cargo for long time periods.

BTS currently occurs beyond their native range. They were introduced to Guam during World War II with a single female snake and spread to all parts of Guam by the late 1960s. Subsequently it has been shown they are responsible for a drastic decline in numbers of all native birds on the island.

6.2.1 Predictive model

The BTS point data were transcribed from an old fan-fold listing of the Australian Museum holdings provided by Gordon Rodda (personal communication 2005) compiled circa 1988 and transcribed to digital form. In all, there were 274 data points covering only the Australian and Papua New Guinean portion of the native distribution.

Whywhere identified the most influential variables and, via study of the response curves, provides insight into way a species responds to its environment. The algorithm uses image processing methods to efficiently sift through large amounts of data to find the few variables that best predict species occurrence.

A habitat suitability map prepared for the BTS using data and the Why-Where model is shown below. The results of mining 528 terrestrial variables was as follows (Figures 6.7).

```
Best accuracy 0.840 for variable lcprc03.raw
Leemans and Cramer March Precipitation (mm/month)
```

It is clear that the response of the BTS is an asymptotic function of precipitation, with no occurrences at low March precipitation, and no real limit

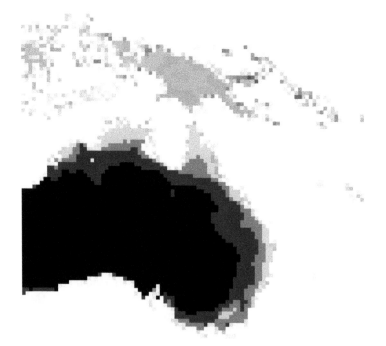

FIGURE 6.6: The distribution of the Brown Treesnake predicted from March precipitation by WhyWhere. Black is zero or low suitability, dark grey is medium and light grey is highly suitable environment.

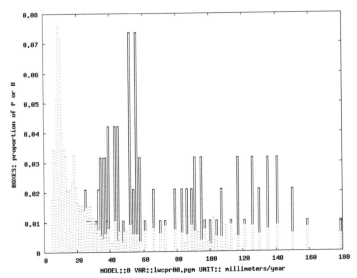

FIGURE 6.7: The histogram of the response of the Brown Treesnake (y axis) to classes of March precipitation (x axis). Dashed bars represent the frequency of the precipitation class in the environment, while solid bars represent the frequency of the BTS occurrences in that precipitation class.

to upper precipitation. This is consistent with observations that BTS prefer a humid environment, and do not inhabit environments with chronic low humidity, and appear unable to shed properly when the relative humidity falls below 60 percent [RFMCI99].

6.3   Invasion of Zebra Mussel

Niche models are generally thought of as equilibrium models – incapable of representing dynamic process such as movement of a species. Here we show one way that the invasion of a species might be modeled.

In previous invasion work, the model of the distribution has been developed from the distribution in the home range of the species. This model is projected onto niche conditions in the new landscape (Figure 6.8). This approach is limited by the availability of adequate points in the home range of the species.

An alternative approach here achieves a prediction of the possible trajectory of the invasion, subject to some assumptions. This makes it potentially more useful, both for predicting possible range of species with no home range data,

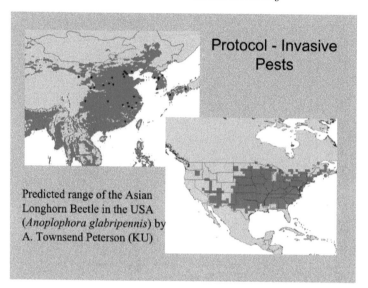

Protocol - Invasive
Pests

Predicted range of the Asian
Longhorn Beetle in the USA
(*Anoplophora glabripennis*) by
A. Townsend Peterson (KU)

FIGURE 6.8:   An effective protocol for predicting the potential distribution
of invasive species is to develop a model on the home range of a species then
predict the distribution using the same environmental variables in the area of
interest.

and in representing the dynamics of the invasion.

The zebra mussel (ZM, *Dreissena polymorpha*) is native to the Black, Caspian,
and Azov seas. The ZM is highly invasive due in part to high rate of fecundity
(e.g., 10,000 to 1,000,000 eggs per female) and a pelagic larval stage, which is
rare among freshwater bivalves.

While the ZM is a recent invader in North American waters, it has been
invading waters of the former Soviet Union, eastern, and western Europe for
over 200 years from its origin in the Black and Caspian Seas in Eurasia. The
initial invasion of the ZM in the Great Lakes area was probably in 1986 or
1987 in southwestern Lake Erie and Lake St. Clair (which are connected by
the Detroit River), presumably when freshwater ballast was discharged by
ocean-crossing vessels from Europe.

In the United States, the ZM has invaded as far south as the lower Missis-
sippi River, although populations at the southern edge of the invasion appear
stressed by summer conditions. A number of habitat parameters, including salinity, calcium concentration, temperature, oxygen concentration, and
substrate, appear to influence ZM distribution. However, due to lack of availability of extensive maps of these variables, we must make do with average
climate variables as predictors.

The approach to defining a dynamic model of the ZM from a static niche

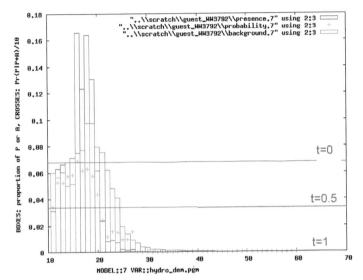

FIGURE 6.9: A simple approach to simulating the spread of an invasive species is to develop a series of predictions by moving a cut value from the peak of the probability distribution to the base.

model is to simulate the spread with a moving cut from the peak of the probability distribution to the base. This process at points is illustrated in Figure 6.9.

This results in a nested sequence of increasing areas, starting with the areas predicted with the highest probability, to the areas of least probability last, as shown on Figure 6.10.

Clearly this approach relies on certain assumptions about the potential invasion. It assumes the areas invaded first have the highest probability of occurrence of the species, while the areas invaded last have the lowest probability. These constraints could be satisfied either in the case of an invasive species that occupied its most favorable habitat first, and spread out from there. Alternatively, the probability distribution itself could incorporate spatial factors correlating probabilities with a possible sequence of invasion, perhaps by using spatial factors in the set of environmental variables.

Figure 6.11 illustrates model accuracy. The crosses are observations of the species, the x axis is the time before present, and the y axis is the cut value. The diagonal line marks the value of the cut at each point in time, creating the simulation. Observations above the diagonal are within the predicted range of the species at a given time, while the observation below the diagonal are outside the predicted range of the species at a given time. In this example, most of the observations lie within the trajectory predicted by the approach.

Niche Modeling

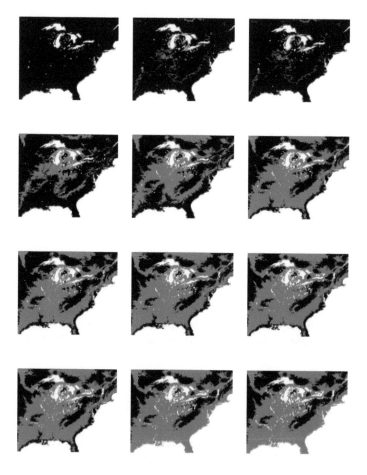

FIGURE 6.10: The nested sequence of predicted ranges, based on movement of the cut value.

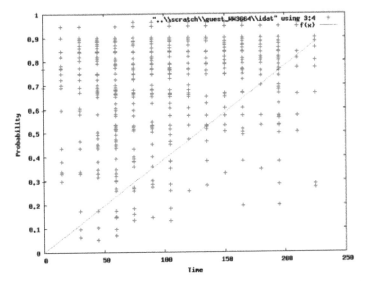

FIGURE 6.11: Evaluation of the accuracy of the prediction of invasion trajectory, with time before present on the x axis and value of cut probability on y axis. Observations above the diagonal are correct predictions, while observations below the diagonal are incorrect predictions.

6.4 Observations

These results illustrate a number of points related to choice of variables. Firstly, as illustrated by the House price increase example, while analysis using climatic variables may give predictive models, consideration of a larger set of variables relevant to the object of interest may lead better models in general.

Secondly, it has been demonstrated that annual climate averages often can be sub-optimal [Sto06]. Here, monthly averages of temperature in general contributed most to accuracy of models. Monthly variables have rarely, if ever, been using in niche modeling, despite potential biological explanation for these variables being important in terms of conditions that limit the range of species.

Thirdly, response to environment is not limited to 'inverted U's'. The house price example showed a bimodal distribution, while the BTS example was asymptotic.

Thus, one must be aware of errors that might result from forcing the model to use previously determined environmental variables and preconceived forms of the species response. Even though a model passes a test of significance it may give sub-optimal results relative to a larger set of variables. The nagging question always hangs over such analysis - were there better variables that would have changed the result?

The approach illustrated by WhyWhere, but equally applicable to other methods is to discover the most accurate variable from among a large set of variables. Thus, claims to model skill have more veracity than claims based on a very limited selection of possible correlates.

Nevertheless, the approach is still vulnerable to many forms of error. The remainder of the book is devoted to explaining some overlooked types of error that can undermine analyses.

Chapter 7

Bias

A bias is generally understood as a prejudice in the sense for having a predilection to one particular point of view or ideological perspective. Statistical bias is a quantifiable tendency of a statistical estimator to over or underestimate the quantity that is being estimated.

In formal terms, if $\hat{\theta}$ is an estimator of a parameter θ, then $\hat{\theta}$ is biased if the expected value of θ, $E(\hat{\theta})$, is not equal to θ.

Bias is a concern when any process is supposed to be objective and fair. Similarly statistical bias of an estimator should be of concern, as that bias can influence results and undermine the objectivity of conclusions.

For example, systematic errors increase the likelihood of assigning statistical significance to chance events (Type I error). Given an hypothesis

$$H_0 : \hat{\theta} = \theta$$

and the alternative

$$H_a : \hat{\theta} \neq \theta$$

then bias would tend to increase the chance of rejecting H_0 when H_0 is true, or committing a Type I error.

Bias is more insidious than uncertainty introduced by small samples. For example, errors in estimation of a mean value are generally symmetric, that is, equally as likely to be above or below the actual value. Thus while a small sample introduces uncertainty, it does not necessarily bias the result.

The detection of bias in modeling should be a primary concern. While increasing the sample size can reduce uncertainty in symmetric errors, averaging is not a solution to asymmetric errors. The variation in symmetric sampling errors reduces our power to detect statistical patterns in data (Type II error). Type II errors only lead us to reserve judgment, but Type I errors cause us to draw incorrect conclusions. In most cases, drawing incorrect conclusions should be a greater concern than failing to find significance.

7.1   Range shift

We look at potential bias in the range-shift methodology. This methodology examines effects of potential shifts in the ranges of species that may be brought on by changes in the environment.

In order to better understand the response of species to environmental change, and potentially predict responses, an ecological niche model is developed based on present day environmental variables and then reapplied using the different variables, either from the past or future. This often leads to a 'shift' in the predicted distribution of the species. Shift-modeling is the name for comparing predicted ranges of species in the past and future with the present.

7.1.1   Example: climate change

The main applications of shift-modeling to date have been the reconstruction of past distributions of species under paleo-climatic conditions, and the possible effects climate change on biological communities.

Climate is always changing. Since the last major extinction at the start of the Pleistocene ice-age at 700kyr BP, the earth's climate system has oscillated between glacial and interglacial states approximately every 200,000 years [PJR$^+$99]. Temperatures in Antarctica varied by about 10°C from peak to trough, and transitions of at least 2°C occur on average every 10,000 years [AMS$^+$97, AMN$^+$03, MGLM$^+$04].

Local climate changes can be even more abrupt. The Younger Dryas event at about 12kyr BP changed local temperatures 7°C in 1000 years, and events at 8.2kyr BP cooled Europe even more abruptly. In response to these changes, the fossil record shows changes to faunal composition and richness with a few recent extinctions attributable to recent climate changes [JW99]. There is some observational evidence that ranges of some fauna and flora have changed in response to the increase of 0.6°C in the last century and will continue to change in response to climate changes [PY03, RPH$^+$03, TWC$^+$04].

7.2   Range-shift Model

Here we develop a simple geometric model to estimate the asymmetric error in simple range-shift niche models. The particular statistic of concern is the change in the potential range area available to species. Note the range-shift methodology does not explicitly represent other factors such as:

- dispersal ability,

- rate of change, or

- threshold effects.

Variation in dispersal ability of individual species in response to change is modeled by two geometric combinations of ranges: new ranges representing free dispersal and the intersection of new and old areas representing no dispersal. These alternatives are intended to 'bracket' the possible range of behaviors.

Simple shift models incorporate rate of change in temperature only by way of dispersal and no-dispersal scenarios. They contain no history effects that could incorporate a threshold effect, and potential novel combinations of climate.

As bias is the tendency for a statistic to over or underestimate the true value, bias is quantified by showing a particular statistic does not estimate the correct value, by comparing the true value with the estimated value. Deviation in the expected value of the new or intersection area due to errors quantifies the bias.

A theoretical model including errors in the parameters allows calculation of both the true area, and the change in area resulting from errors in those parameters. Range area predictions are regarded as biased if the observed value of range area A_e does not equal the expected value A_e. Thus area can be treated as a statistical parameter like accuracy [SP02a].

Consider a shape such as a circle with width, or radius r. Shifting the shape laterally by s produces areas:

- old A_O,

- new A_N,

- an intersection area A_I, and

- a union A_U.

Niche Modeling

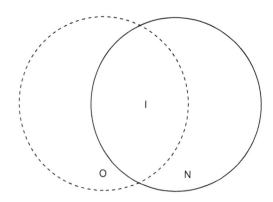

FIGURE 7.1: Theoretical model of shift in species distribution from change in climate. Dashed circle marked O is old range, solid circle marked N is new range and I is intersection area.

These areas can be calculated for squares and circles using the usual formulas. In this case the area of the square is $A_O = (2r)^2$. The formula for intersection areas of a circle for shift s when radius r remains a constant value of 1 is:

$$A_I = 2r^2 cos - 1(s^2/2sr)$$

The equation for intersection of a square is somewhat simpler:

$$A_I = 2r(2r - s)$$

Using the formulation of the square leads to simpler calculations. Figure 7.2 shows the intersection areas for squares and circles at different shifts for squares (solid line) and for circles (dashed line). The intersection area of a square is only slightly greater than the intersection area of the circle. We will consider only the square in order to simplify calculations.

We now incorporate three error terms into the equations:

- error in the shift term s_e,

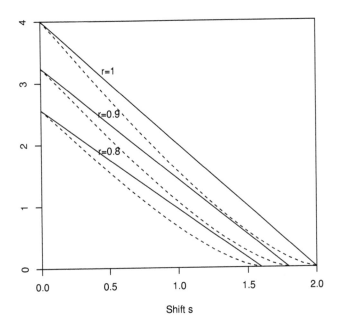

FIGURE 7.2: The change in the areas of intersection of a square and circle for different shifts (s) and widths (r).

- error in width term r_e, and

- error in the proportional change in area A_N/A_O or p_e.

The new area is given by:

$$A_N = (2rr_e)^2 p_e$$

The intersection area is given by:

$$A_I = 2rr_e(2rr_e - ss_e)p_e$$

7.3   Forms of bias

The model quantifies the reduction in intersection area simply as a function of the geometric considerations. A number of consequences are immediately apparent:

- With no shift or width error, shift alone will not reduce ranges.

- For a given shift, species with smaller ranges will have greater reduction in intersection areas that species with larger ranges.

- Due to edge effects, proportional areal reduction depends on the size of shifts relative to the size of the study region.

- Apparent new ranges are considerably reduced by width and shift errors that reduce new widths and increase apparent shifts.

We now look at motivations for size and direction of the individual error terms.

7.3.1   Width r and width error

Generally it is assumed that width as determined by the model of occurrences is a good indication of the actual range of the species. However model determined width may be subject to errors. Researchers have noted the problem of recovering accurate response curves from incomplete geographic

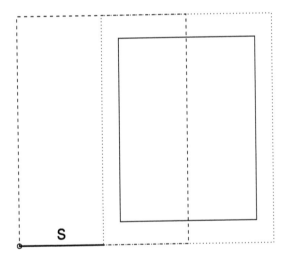

FIGURE 7.3: Combined effect of shift and width error.

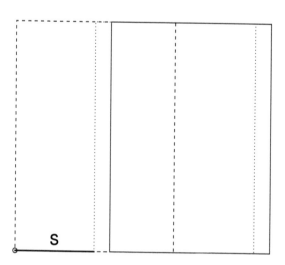

FIGURE 7.4: Combined effect of shift and shift error

ranges [AM96, TBAL04]). Because the samples of observations are incomplete, all empirical models underestimate width of the potential niche, or 'overfit' [SP02b]. Thus there will inevitably be areas of potential habitat outside the known observations.

The consequence of overfitting is prediction of a smaller area than would otherwise be predicted. Overfitting has been shown to be a potent source of bias in climate envelope methods due to the truncation of the range of variables as discussed previously, underestimates of areas by up to 15% [BHP05]. Other studies have shown similar errors in range due to sample size. Where overfitting is defined as the difference between accuracy of a model on a training set versus a test set, errors range from 30% for 10 data points to around 5% at 100 points when datasets are sufficiently large [SP02b]. Studies indicate overfitting also varies by method. In one study of a large number of bird species, overfitting of species with twenty observations was approximately 10% with GARP and 15% for logistic regression [SP02b]. Figure 7.3 illustrates the combined effect of shifting the square range (dotted) and width error (solid).

7.3.2 Shift *s* and shift error

A major source of shift error is differential response of species to climate change. It is generally assumed that the only two possible responses to climate change are migration to new ranges or contraction of the range to the intersection of the old and the new area.

If, however, some species maintain old ranges or expand to form a union of old and new ranges, average areal change is overestimated. This could come about if climate sensitivity was orthogonal to dispersiveness. Some physiological studies show climate based ecological niche models exaggerate the effects of climate, ignoring complex responses of species to other factors [Ll96, SCF+01, Loe03]. Differential (non-climatically driven) behavior in the leading and trailing edges of ranges is supported by observational evidence [PY03, RPH+03, TTR+04]. For example, many temperate trees can survive outside their current range as sub-dominants even though dominated by faster growing sub-tropicals [Loe03].

7.3.3 Proportional p_e

Map boundaries are a potential source of width bias in range-shift modeling. As climate patterns shift, a certain percentage of distributions will be moved out of study area, reducing their apparent size. Compensation cannot occur elsewhere in the study area because: (1) the ranges of species entirely outside the old area are not represented, and (2) truncated distributions are fit to species at edges. This is an artifactual effect introduced because species' ranges cover the landscape beyond the limited study area.

7.4 Quantifying bias

Figure 7.5 shows the effect of new range from all errors. We illustrate the effects of these errors in the parameters by substituting the following values into the theoretical shift model.

- unit radius $r = 1$,

- radius error $r_e = 0.85$,

- shift $s = 0.8$,

- shift error $s_e = 1.33$ and

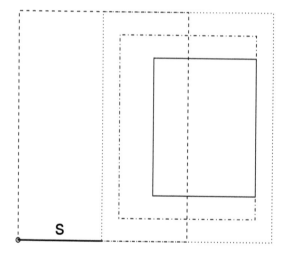

FIGURE 7.5: Combined effect of shift, shift error, width error and proportional error.

- proportional change error $p_e = 0.85$

Substitution of these parameters into the theoretical model gives the following results.

- With no bias errors the new area is an unchanged 4 and the intersection area is 2.4.

- All errors combined reduce new area to 2.46 and intersection area to 0.96.

- The theoretical model suggests errors from biases reduce the size of new areas to 62% and the size of intersections areas to 40% of their expected values.

7.5 Summary

The theoretical range-shift model estimates bias in the range-shift method. The bias results in exaggerated estimates of reduction in range, and potentially exaggeration of risk to species from climate change. Strong and unambiguous conclusions are not justified when there is significant biases, and can be a considerable source of errors.

Chapter 8

Autocorrelation

Correlation indicates a relationship between two variables. In simple terms, when one 'wiggles' the other 'wiggles' too. In autocorrelation, instead of correlation between two different variables, the correlation is between two values of the same variable at different times or different places.

The autocorrelation function (ACF) of a variable X describes the correlation at different points X_i and X_j. If X has a mean of μ and variance of σ^2 the ACF as a function of two points i and j where E is the expected value is given by:

$$ACF(i,j) = \frac{E[(X_i - \mu)(X_j - \mu)]}{\sigma^2}$$

Autocorrelation occurs in both the spatial context of environmental variables and the temporal context of time series analysis.

The main concern with autocorrelation is that failing to take it into account can produce exaggeration of significance and hence errors, e.g.:

> Correlation between an autocorrelated response variable and each of a set of explanatory variables is highly biased in favor of those explanatory variables that are highly autocorrelated [Len00].

That is, multiple regression will find a variable with high autocorrelation 'significant' more often than it should, and therefore be featured more highly in a model than it deserves, possibly replacing a best variable without autocorrelation. It has been claimed that models niche models may introduce 'low frequency' variables like temperature and rainfall falsely into models due to the high autocorrelation in climate variables. In a fair comparison, 'high frequency' variables such as vegetation could be as accurate or better [Len00].

It is important therefore for successful niche modeling to understand autocorrelation and how it can lead to errors. The simplest way to study and understand autocorrelation is to look at the one dimensional case of time series, rather than 2D to which most results generalize.

Here we construct a set of the basic types of series to examine their properties.

8.1   Types

While basic features such as the mean, standard deviation and linear trends are usually the basis of analysis, little attention is usually paid to the auto-correlation properties of these models.

There are a number of ways of generating autocorrelation. These internal features also have a bearing on explanations for phenomena.

As an example, we determine the parameters for different types of series matching the parameters derived from global temperature. We use the global temperatures from the mid-nineteenth century to the present recorded by the Climate Research Unit (CRU) [Uni].

8.1.1   Independent identically distributed (IID)

An IID series is the simplest and most familiar series consisting of independent random numbers with a distribution such as the normal distribution. Future terms in the series are determined by the long term mean a and variance of past data. Specifically, each value is not dependent on any other term. For example where e is a normally distributed random variable:

$$X_t = e$$

The series of random numbers with a normal distribution and a standard deviation equal to CRU data is shown in Figure 8.1.

8.1.2   Moving average models (MA)

In moving averages, the average of a limited set or window of values is calculated at every position in the series. In R this is done with the filter command, the filter being determined by a list of numbers to use as coefficients in a summation – in this case 30 values of $1/30$ provide a 30 year moving average for CRU. A MA is often called a low frequency band pass filter, as it suppresses high frequency fluctuations while passing the long frequency ones. Here is an equation for generating a moving average shown in Figure 8.1:

$$X_t = \sum_{i=1}^{n} \frac{X_{t-i}+e}{n}$$

8.1.3 Autoregressive models (AR)

In auto-regression models each term in the series is determined by the previous terms plus some random error. In an AR(1) (or Markov) model only the previous term is used in predicting the next term.

Each term in the AR(1) series where a is a coefficient and e is a random error term can be generated from the following equation

$$X_t = e + aX_{t-1}$$

A random walk is a form where $a = 1$. A walk can be generated from a series of random numbers by taking the cumulative sum.

We can estimate the value of a in R with the $ar()$ function and the CRU temperature data. We can then generate an AR(1) model using the R facility arima.sim with the given parameters. The coefficient is $a = 0.67$ and standard deviation is $sd = 0.15$ for the AR(1) model of CRU.

8.1.4 Self-similar series (SSS)

The next series goes by many names: self-similar, fractal, roughness, fractional Gaussian noise model (FGN), long term persistence (LTP), clustering or simple scaling series (SSS). Mostly they are characterized as having constantly scaling variance (or standard deviation) over all time or spatial scales, and hence the term simple scaling series is most accurate. Fractional differencing, is a generalization of integer difference series, where the degree of differencing is allowed to take any real value rather than being restricted to integers.

For example, in normal Brownian motion, the value of a series X_t at time t is dependent on its previous value X_{t-1} and the random variable a has a difference of one. In the following X_t is a function of the partial sum of all terms preceding it. The integer differencing operator is written in terms of a backshift operator B as:

$$(1 - B)X_t = at$$

The fraction difference operator $(1 - B)^d$ is defined by the binomial series where kth term in the series is summed from 0 to infinity, and d is a function of the Hurst exponent $d = H - 0.5$. These are called FARIMA models. A $FARIMA(0, d, 0)$ process is written:

$$(1 - B)^d X_t = \sum_{k=0}^{\infty} \binom{d}{k} - B^k$$

R has a package called fracdiff that allows estimation of the parameters of ar, d, and ma for simulation of a $FARIMA(ar, d, ma)$ process where ar and ma are the classical $ARMA(ar, ma)$ parameters.

8.2   Characteristics

In Figure 8.1 the simulated series are plotted. The AR(1) and the SSS series resemble quite closely the CRU natural series. However the IID series does not capture the longer time scale fluctuations. In comparison, the random walk is difficult to plot as it tends to trend so strongly it walks out of the figure area.

While it can be seen by eye in Figure 8.1 that IID and random walk are not good models for the natural series more insightful methods are needed to distinguish them. Highly autocorrelated models are described as having 'fat tails'. This refers to the way the distribution of less frequent difference values fades out into a thicker tail (power-type) rather than the exponential form of a normal distribution. When these distributions are plotted in Figure 8.2 it is hard to see which are power and which are not. We need more powerful ways to examine the data.

8.2.1   Autocorrelation Function (ACF)

One of the main tools for examining the autocorrelation structure of data is the autocorrelation function or ACF. The ACF provides a set of correlations for each distance between numbers in the series, or lags. The autocorrelation decays in a characteristic fashion for each series as the lags get longer as shown in Figure 8.3. It can be seen that the autocorrelations of the IID series decay very quickly (no long term correlation), the AR(1) model decays fairly quickly, the SSS next and the random walk most slowly.

The characteristic decay in autocorrelations relative to the inverse and inverse log plot is sometimes easily seen by plotting the log of the y axis (Figure 8.4).

A second tool for examining the autocorrelation structure of data is the lag plot. Figure 8.5 shows the autocorrelated processes CRU, CRU30, AR1.67, WALK and SSS with diagonals, while the random IID variable is a cloud of points. Smoothing greatly increases the diagonalization of the points on the

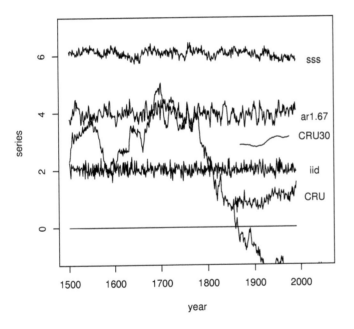

FIGURE 8.1: Plots of the global temperatures (CRU), the simulated series random, walk, ar(1), and sss.

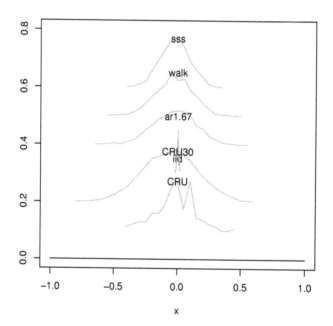

FIGURE 8.2: Probability distributions for the differenced variables.

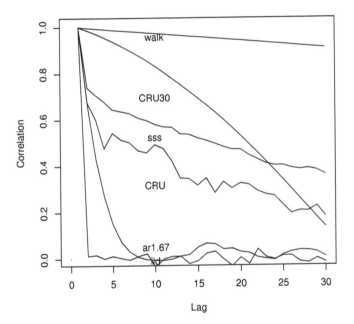

FIGURE 8.3: Autocorrelation function (ACF) of the simulated series, with decay in correlation plotted as lines. Degree of autocorrelation is readily seen from the rate of decay and compared with temperatures (CRU).

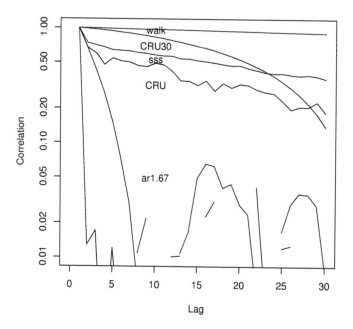

FIGURE 8.4: Highly autocorrelated series are more clearly shown when plotting on a log plot. The IID and simple Markov AR1.67 series decline most rapidly. Note also that the autocorrelation of the moving average of CRU temperatures tends to decline more rapidly than the raw CRU series.

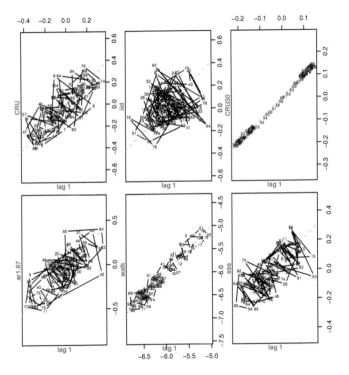

FIGURE 8.5: Lag plot of the processes CRU, IID, CRU30, AR1.67, walk, and SSS. Autocorrelated series exhibit strong diagonals.

lag plot in CRU30.

8.2.2   The problems of autocorrelation

The previous figures illustrated that while ARMA series have some of the autocorrelation properties required for simulating climatic series, but do not generate sufficient long tem correlations [Kou02] to represent natural series data. For this, fractional differencing of the simple scaling series was needed. Thus, representation of the autocorrelation properties of natural series is not possible with the majority of simple IID or AR(1) models in use.

The problems of autocorrelation stem from the difficulty of adequate validation of the significance of results. Even if a model is validated on data points 'held back' from the model calibration, autocorrelation will result in overestimates of significance. This is because the degree of independence varies according to separation of points, and it is sometimes impossible to entirely separate the validation period from the calibration period. Thus it can be difficult to obtain truly independent tests of a model.

We illustrate this effect using two different statistical measures: the r^2 statistic and the reduction-of-error or RE statistic applied to a simple model of temperatures with autocorrelation.

The r^2 statistic, also called the Coefficient of Determination, is widely used in regression models to indicate the degree of correlation of the independent to the predicted values. It is calculated from SSE the sum of squares of the errors and SSM the sum of squares of the mean.

$$r^2 = 1 - SSE/SSM$$

The r^2 can be either positive in a positive correlation or negative in a negative correlation. An indication of skill the r_2 will be positive value, the closer to one the better.

The RE statistic is as follows, where x are the actual values and y are the predicted values.

$$RE = 1 - \frac{\sum (x-y)_2}{\sum (x-\bar{x})^2}$$

RE can be negative or positive, but a positive value generally indicates skill. The RE is positive if the model-predicted values are somewhat better predictions than the mean value. Unlike the r^2 statistic which is independent of differences in magnitude of the two series being correlated, the RE penalizes the predicted values for deviation from the mean value.

8.3   Example: Testing statistical skill

Here we determine the skill of a very simple Monte Carlo model that should have no skill at all in predicting values outside those points used for calibrating the model.

The model predictions shown as a black line in Figure 8.6 were created by generating series at random and selecting those that correlated by the r^2 statistic with CRU temperatures. Those 20% that correlated were then averaged.

The full procedure is as follows:

- split CRU temperature data into a training set and a test set,

- generate 100 random SSS sequences of length 2000 (years),

- select those sequences with positive slope and $r^2 > 0.1$ on training data,

- calibrate the sequence using the inverse of the linear model fit to that sequence,

- smooth the sequences using a 50 year Gaussian filter,

- average the sequences, and

- calculate the r^2 and RE statistic against raw and filtered temperature data on training and test sets.

The averaging of these correlated series results in a reconstruction of temperatures, both within and outside the calibration range. While one expects this model to fit the data within the calibration range, one expects the model to have no skill at predicting outside that range as it is based on entirely random numbers.

Here we are comparing the results achieved from comparing a number of alternatives:

- correlations on the test period, then on the independent period,

- comparing different ways the independent sample can be drawn,

- comparing the different statistics, r^2 and RE, and

- comparing raw data versus smoothed data.

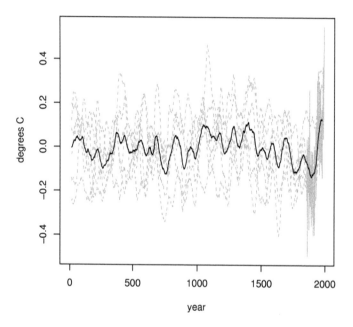

FIGURE 8.6: As reconstruction of past temperatures generated by averaging random series that correlate with CRU temperature during the period 1850 to 2000.

8.4 Within range

In the cross-validation procedure, the test data (validation) are selected at random from the years in the temperature series in the same proportion as the previous test. Those selected data are deleted from the temperatures in the training set (calibration). This represents a fairly typical 'hold back' procedure.

The results of r^2 and RE are shown in the Table 8.1 below. Both statistics appear to indicate skill for the reconstruction on the within-range calibration data. However, both statistics also indicate skill on the cross-validation test data using significance levels RE>0, r^2>0.2. These results are significant both for the raw and smoothed data.

TABLE 8.1: Both r^2 and RE statistics erroneously indicate skill of the random model on in-range data.

	Period	R2	s.d.	RE	s.d..1
1	Training	0.50	0.07	0.17	0.34
2	Test	0.51	0.09	0.22	0.34
3	Training smooth	0.86	0.06	0.25	0.37
4	Test smooth	0.87	0.09	0.26	0.35

8.4.1 Beyond range

In the following test the RE and the r^2 statistics for the reconstruction from the random series is calculated on beyond-range temporally separate test and training periods. The period is at the beginning of the CRU temperature record, marked with a horizontal dashed line (test period to the left, training to right of temperature values).

Most of the r^2 values for the individual series however are below the cutoff value of 0.2 on the test set which can be regarded as not significant. This indicates by cross-validation statistics the model generated with random sequences have no skill at predicting temperatures outside the calibration interval.

However, Table 8.2 below shows the r^2 for the smoothed version, still indicate significant correlation with the beyond-range points. This illustrates the effect of high autocorrelation introduced by smoothed temperature data (a highly significant 0.44). It could be erroneously claimed that the smoothed reconstruction has skill at predicting the temperatures in the beyond-range period.

The table below adds the RE statistic to the previous RE statistic. The

TABLE 8.2: While r^2 indicates skill on the smoothed model on out-of-range data, RE indicates little skill for the random model.

	Period	R2	RE
1	Training	0.56	0.40
2	Test	0.02	−0.35
3	Training smooth	0.91	0.45
4	Test smooth	0.44	−0.39

results for RE are positive in comparisons of the reconstruction over training data, but are negative for comparisons on the test data. In this case, RE statistics would not validate the skill of the reconstruction for predicting temperatures either on the raw or smoothed model.

8.5   Generalization to 2D

These results of the one dimensional example apply to predictions of species distributions in 2D. The accuracy achieved with randomly selected species occurrences within the range of species may be a poor indicator of the accuracy of the model beyond that range. Equivalently, the only reliable indication of the accuracy of a niche model may be the accuracy on areas extremely distant from the area where the model is calibrated. For example, the only real test of a niche model might be the capacity to predict the range of an invasive species on a new content. However, an invasion is mediated by other factors can affect determination of prediction accuracy in this way.

8.6 Summary

These results demonstrate the problems of validating models on autocorrelated data. While this is only one approach to cross-validation on held-back data, it shows that data randomly selected within an autocorrelated series will not adequately determine model skill. Different statistics can also give different results, as shown by the r^2 and RE cross-validation statistics.

In fact, as shown with smoothed data, even tests applied to held back data out of range of the calibration period may still appear to be significant. A great deal of care needs to be taken to ensure validation tests are sufficiently severe to detect unskillful models with autocorrelated data.

Chapter 9

Non-linearity

In mathematics, non-linear systems are, obviously, not linear. Linear system and some non-linear systems are easily solvable as they are expressible as a sum of their parts. Desirable assumptions and approximations flow from particular model forms, like linearity and separability, allowing for easier computation of results.

In the context of niche modeling, an equation describing the response of a species y given descriptors $x_1, ... x_n$ is obviously linear:

$$y = a_1 x_1 + ... + a_n x_n \qquad (9.1)$$

However this form of equation is unsuitable for niche modeling as it cannot represent the inverted 'U' response that is a minimal requirement for representing environmental preferences. The second order polynomial can represent the curved response appropriately and is also linear and easily solvable:

$$y = a_{11} x_1 + a_{12} x_1^2 + ... + a_{n2} x_n^2$$

However, our concern here is not with solvability. Our concerns are the errors that occur when non-linear (i.e. curved) systems such as equation 9.2 above, are modeled as linear, systems like equation 9.1.

Specifically we would like to apply niche modeling methods and determine how types of non-linearity affect reliability of models for reconstructing past temperatures over the last thousand years from measurements of tree ring width. This will demonstrate another potential application of niche modeling to dendroclimatology. Response of an individual and species as a whole to their environment is basic to climate reconstructions from proxies. Simulation, first in one and then two dimensions, can help to understand the potential errors in this methodology from non-linearity.

9.1　Growth niches

While we use a one dimension example of reconstructions of temperatures for simplicity, the results are equally applicable in two dimensions. However, we initially use actual reconstructions of past temperature.

Reconstructions of past climates using tree-rings have been used in many fields, including climate change and biodiversity [Ker05]. It is believed by many that

> "carefully selected tree-ring chronologies ... can preserve such co-herent, large-scale, multicentennial temperature trends if proper methods of analysis are used" [ECS02].

The general methodological approach in dendroclimatology is to normalize across the length and the variance of the raw chronology to reduce extravagant juvenile growth, calibrate a model on the approximately 150 years of instrument records (climate principals), and then apply the model to the historic proxy records to recover past temperatures. The attraction of this process is that past climates can then be extrapolated back potentially thousands of years.

Non-linearity of response has not been greatly studied. Evidence for nonlinear response emerges from detailed latitudinal studies of the responses of single species to multiple climate principals [LLK+00]. Mild forms of non-linearity such as signal saturation were mentioned in a comprehensive review of climate proxy studies [SB03]. These are described variously as a breakdown, 'insensitivity', a threshold, or growth suppression at higher temperature. Here we explore the consequences of assuming the response is linear when the various forms of growth response to temperature could be:

- linear,

- sigmoid,

- quadratic (or inverted 'U'), and

- cubic.

Such non-linear models represent the full range of growth responses based on knowledge of the species physiological and ecological responses, is basic to niche modeling, and a logical necessity of upper and lower limits to organism survival.

TABLE 9.1: Global temperatures and
temperature reconstructions.

	names	Reference
1	year	year
2	CRU	Climate Research Unit
3	J98	Jones et al. 1998 Holocene
4	MBH99	Mann et al. 1999 Geophys Res Lett
5	MJ03	Mann and Jones 2003
6	CL00	Crowley and Lowery 2000 Ambio
7	BJ00	Briffa 2000 Quat Sci Rev
8	BJ01	Briffa et al. 2001 J Geophys Res
9	Esp02	Esper 2002 Science
10	Mob05	Moberg 2005 Science

The series we examine in the linear and sigmoidal sections are listed in
Table 9.1. Non-linear models for the quadratic reconstructions were constructed from an ARMA time series of length 1000 with the addition of a
sinusoidal curve to approximate the temperatures from the Medieval Warm
Period (MWP) at around 1000AD through the Little Ice Age (LIA), to the
period of current relative warmth. The coefficients were determined by fitting
an ARIMA(1,0,1) model to the residuals of a linear fit to 150 annual mean
temperature anomalies from the Climate Research Unit [Uni] resulting in the
following coefficients: AR=0.93, MA = -0.57 and SD = 0.134.

9.1.1 Linear

A linear model is fit to the calibration period using $r = at + c$. The linear
equation can be inverted to $t = (r - b)/a$ to predict temperature over the
range of the proxy response as shown in Figure 9.1. Some result in better
reconstructions of themselves than others, depending largely on the degree of
correlation with CRU temperatures over the calibration period.

The table 9.2 shows slope and r_2 values for each reconstruction. All reconstructions have generally lower slope than one. While a perfect proxy of
temperature would be expected to have a slope of one with actual temperatures, this loss of sensitivity might result from inevitable loss of information
due to noise [vSZJ$^+$04].

9.1.2 Sigmoidal

It is of course not possible for tree growth to increase indefinitely with
temperature increases; it has to be limited. The obvious choice for a more

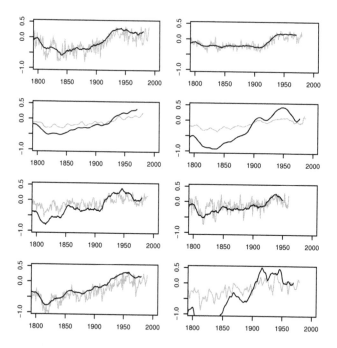

FIGURE 9.1: Reconstructed smoothed temperatures against proxy values for eight major reconstructions.

TABLE 9.2: Slope and correllation coefficient of temperature reconstructions with temperature.

	Slope	r2
J98	0.78	0.35
MBH99	0.84	0.69
MJ03	0.43	0.47
CL00	0.27	0.23
BJ00	0.36	0.24
BJ01	0.57	0.33
Esp02	0.80	0.41
Mob05	0.28	0.09

accurate model of tree response is a sigmoidal curve. To evaluate the potential of a sigmoidal response I fit a logistic curve to each of the studies and compared the results with a linear fit on the period for which there are values of both temperature and the proxy. The results were as follows (Figure 9.2).

The logistic curve did not give a stunning increase in the r^2 valuesÃć although they were comparable. I had to estimate the maximum and minimum temperatures for each proxy from the maximum value and 0.1 minus the minimum value. Perhaps there is room for improvement in estimating these parameters as well and would improve the r^2 statistic.

9.1.3 Quadratic

The possibility of inverted U's in the proxy response is even more critical with possibility that growth suppression at higher temperatures may have happened in the past. Figure 9.3 shows an idealized tree-ring record, with a linear calibration model (C) and the reconstruction resulting from back extrapolation. Due to the fit of model to an increasing proxy, smaller rings indicate cooler temperatures. A second possible solution (dashed) due to higher temperatures is shown above. Thus the potential for smaller tree-rings due to excess heat, not excess cold, affects the reliability of specific reconstructed temperatures after the first return to the maximum of the chronology. It is obvious that in this case statistical tests on the limited calibration period will not detect nonlinearity outside the calibration period and will not guarantee reliability of the reconstruction.

The simple second order quadratic function for tree response to a single

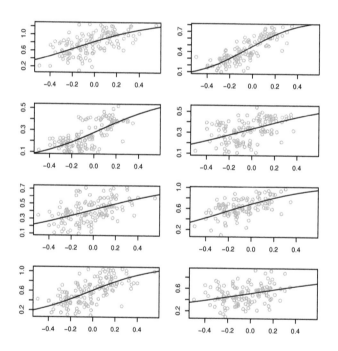

FIGURE 9.2:   Fit of a logistic curve to each of the studies.

climate principle is:

$$r = f(x) = max[0, ax^2 + bx + c)]$$

Where addition of a second climate principle is necessary, such as precipitation, the principle of the maximum limiting factor is a simple and conventional way to incorporate both factors. Here $t \epsilon T$ is temperature and $p \epsilon P$ is precipitation (Figure 2C):

$$r = min[f(t), f(p)]$$

The inverse function of the quadratic has two solutions, positive and negative:

$$t = -b \pm \sqrt{\frac{b^2 - 4a(c-r)}{2a}}$$

This formulation of the problem clearly shows the inherent nondeterminism in the solution, producing two distinct reconstructions of past climate. In contrast, the solution to the linear model produces a single solution.

Figure 9.4 shows the theoretical quadratic growth response to a periodically fluctuating environmental principle such as temperature. The solid lines are the growth responses for three trees located above, centered and below their optimum temperature range. The response has two peaks, because the optimum temperature is visited twice for a single cycle in temperature. Note the peaks are coincident with the optimal response temperatures, not the maximum temperatures of the principle (dotted lines). The peak size and locations do not match the underlying temperature trend.

Figure 9.5 shows the theoretical growth response to two slightly out of phase environmental drivers, e.g. temperature and rainfall) where the response function is the limiting factor. The resulting response now has four peaks, and the non-linear response produces a complex pattern of fluctuations centered on the average of climate principles. The addition of more out of phase drivers would add further complexity.

To describe this behavior in niche modeling terms, a tree has a preference function determined by climatic averages and optimal growth conditions. When the variation in climate is small, e.g. the climate principle varies only within the range of the calibration period, the signal is passed unchanged from principal to proxy. But when the amplitude of variation is large, as in the case for extraction of long time-scale climate variation, the amplitude of the proxy is limited, and the interpretation becomes ambiguous.

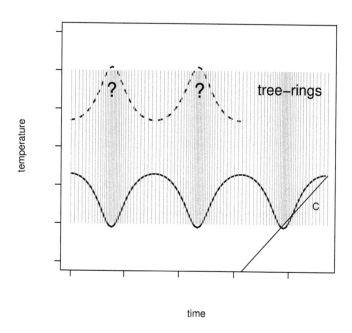

FIGURE 9.3: Idealized chronology showing tree-rings and the two possible solutions due to non-linear response of the principle (solid and dashed line) after calibration on the end region marked C.

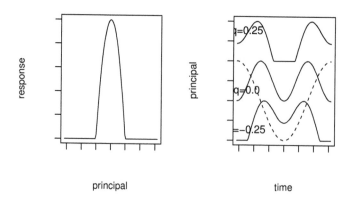

FIGURE 9.4: Nonlinear growth response to a simple sinusoidal driver (e.g. temperature) at three optimal response points (dashed lines).

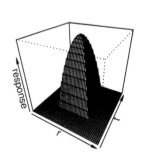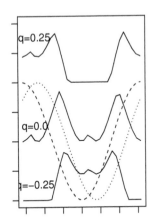

FIGURE 9.5: Nonlinear growth response to two out of phase simple sinusoidal drivers (e.g. temperature and rainfall) at three response points. Solid and dashed lines are climate principles; dotted lines the response of the proxies.

We now examine the consequences of reconstructing temperatures from non-linear responses calibrated by four different regions of the response curve (models R1-4). The left-hand boxes in Figures 9.7, 9.8, 9.9 and 9.10 illustrate the linear and nonlinear calibration models (lines) from the subset years of the series (circles). The right-hand graphs show the reconstructions (solid line) resulting from inversion of the derived model plotted over the temperatures (dots).

9.1.3.1 R1

The first model is a linear model fit to the portion of the graph from 650 to 700 (Figure 9.7). This corresponds to a reconstruction using the portion of the instrumental record where proxies are responding almost linearly to temperature. While the model shows a good fit to the calibration data, and the reconstruction shows good agreement over the calibration period, the prediction becomes rapidly more inaccurate in the future and the past. The amplitude of the reconstructed temperatures is 50% of the actual temperatures, and the peaks bear no relation to peaks in temperature.

Temporal shift in peaks may be partly responsible for significant offsets in timing of warmth in different regions [CTLT00]. The nonlinear model suggests that timing of peaks should be correlated with regional location of trees, and may be a factor contributing to apparent large latitudinal and regional variations in the magnitude of the MWP [MHCE02].

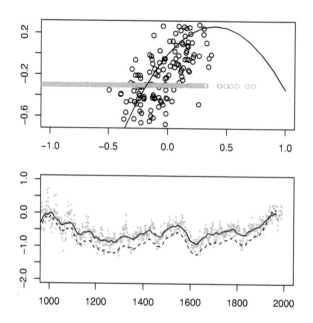

FIGURE 9.6: Example of fitting a quadratic model of response to a reconstruction. As response over the given range is fairly linear, reconstruction does not differ greatly.

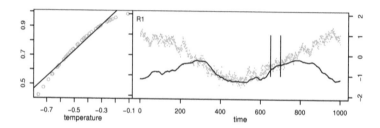

FIGURE 9.7: Reconstruction from a linear model fit to the portion of the graph from 650 to 700.

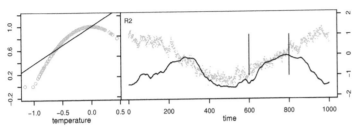

FIGURE 9.8: A linear model fit to years 600 to 800 where the proxies show a significant downturn in growth.

We have also shown here that linear models can either reduce or exaggerate variance. By visual inspection of Figure 9.7, it can be seen that ignoring one of the two possible solutions reduces the apparent amplitude of the long time-scale climate reconstructions by half.

9.1.3.2   R2

The second row in Figure 9.8 is a linear model fit to years 600 to 800 corresponding to reconstruction practice where the proxies show a significant downturn in growth. Model R2 simulates the condition we may be entering at present, described as increasing 'insensitivity' to temperature, as temperatures pass over the peak of response and lower the slope [Bri00]. Another study of tree-line species in Alaska attributed a significant inverted U-shaped relationship between the chronology and summer temperatures to a negative growth effect when temperatures warm beyond a physiological threshold [DKD+04].

The position of the peaks is the same as in the previous models, but due to the inversion of the linear model, the amplitude of the temperature reconstruction is greatly increased. In contrast to R1, the inverted function overestimates the amplitude of past temperatures. Figure 9.8 shows variance and may be exaggerated as calibrated slope decreases.

9.1.3.3   R3

The third, Figure 9.9, is a quadratic model derived from data years 700 to 800, which corresponds to a record of the period of ideal nonlinear response to the driving variable. The resulting reconstructions shows the accurate location and amplitude of the peaks, up to the essential nondeterminism. This demonstrates that an accurate reconstruction of temperature could potentially be recovered if it were possible to choose the correct solution at each inflection point.

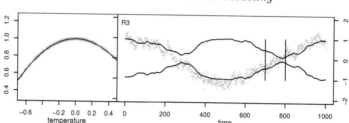

FIGURE 9.9: Reconstruction from a quadratic model derived from data years 700 to 800, the period of ideal nonlinear response to the driving variable.

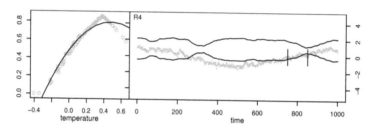

FIGURE 9.10: Reconstruction resulting from a quadratic model calibrated from 750 to 850 with two out of phase driving variables, as shown in Figure 9.5

9.1.3.4   R4

The fourth model is a quadratic model corresponding to a reconstruction fit to the period of maximum response of the species from 750 to 850 with two out of phase driving variables, as shown in Figure 9.10. The poor reconstruction shows that confounding variables together with nondeterminism greatly increase the difficulty of accurately reconstructing climate in greater than one nonlinear dimension.

Lindholm *et al.* [LLK+00] show that while trees at high latitude show strong positive correlation with temperature, trees at middle latitudes show a very confused response. The nonlinear model developed here is consistent with high latitude trees corresponding more closely to temperature, and trees in the middle of the range being more disrupted by the two, temperature and rainfall principles. Thus niche modeling demonstrates its usefulness as a theory for explaining aspects of ecology not previously explained in the linear model.

9.1.4   Cubic

A research project to insert salient features of 'normal' nonlinear physiological behavior between the proxy and principle could go in a number of directions. One could draw from plant ecology where probability of occurrence of species across a temperature ecotone is well modeled by a skew quadratic [Aus87], incorporate the tendency to flattened peaks suggested both by comparison of ring width and density indices, or even asymptotic behavior suggested by physiological models of temperature response of high latitude trees [Loe00].

A second order quadratic model of responses to temperature has the advantage of solvability. However, the simple quadratic response is not proposed as an exact theory. Rather, it is used to provide insights into the expected behavior of the modeled system under ideal or average conditions, and to develop intuition for the consequences of nonlinearity.

9.2 Summary

These results demonstrate that procedures with linear assumptions are unreliable when applied to the non-linear responses of niche models. Reliability of reconstruction of past climates depends, at minimum, on the correct specification of a model of response that holds over the whole range of the proxy, not just the calibration period. Use of a linear model of non-linear response can cause apparent growth decline with higher temperatures, signal degradation with latitudinal variation, temporal shifts in peaks, period doubling, and depressed long time-scale amplitude.

Chapter 10

Long term persistence

Below is an investigation of scaling or long term persistence (LTP) in time series including temperature, precipitation and tree-ring proxies. The recognition, quantification and implications for analysis are drawn largely from Koutsoyiannis [Kou02]. They are characterized in many ways, as having long memory, self similarity in distribution, 'long' or 'fat' tails in the distribution or other properties.

There are important distinctions to make between short term persistence (STP), and LTP phenomena. STP occurs for example in Markov or AR(1) process where each value depends only on the previous step. As shown previously, the autocorrelations in an STP series decay much more rapidly than LTP. In addition, LTP are related to a number of properties that are interesting in themselves. These properties may not all be present in a particular situation, and definitions of LTP also vary between authors. Some of the properties are [KMF04] :

Defn. I **Persistent autocorrelation at long lags:** Where $\rho(k)$ is the autocorrelation function (ACF) with lag k then a series X_t is LTP if there is a real number $\alpha \in (0, 1)$ and a constant $c_p > 0$ such that

$$\lim_{k \to \infty} \frac{\rho(k)}{c_p k^{-\alpha}} = 1$$

In other words, the definition states that the ACF decays to zero with a hyperbolic rate of approximately $k^{-\alpha}$. In contrast the ACF of a STP process decays exponentially.

Defn. II **Infinite ACF sum:** As a consequence of the hyperbolic rate of decay, the ACF of a LTP is usually non-summable:

$$\sum_k \rho(k) = \infty$$

Defn. III **High standard errors for large samples:** The standard error, or variances of the sample mean of a LTP process decay more slowly than the reciprocal of the sample size.

$VAR[X^m] \sim a_2 m^{-\alpha}$ as $m \to \infty$ where $\alpha < 1$

Here m refers to the size of the aggregated process, i.e. the sequential sum of m terms of X. Due to this property, classical statistical tests are incorrect, and confidence intervals underestimated.

Defn. IV **Infinite power at zero wavelength:** The spectral frequency obeys an increasing power law near the origin, i.e.

$$f(\lambda) \sim a\lambda^{\frac{-\alpha}{2}}$$

as wavelength $\lambda \to 0$. In contrast, with STP $f(\lambda)$ at $\lambda = 0$ is positive and finite.

Defn. V **Constant slope on log-log plot** The rescaled adjusted range statistic is characterized with a power exponent H

$$E[R(m)/S(m)] \sim am^H \text{ as } \to \infty \text{ with } 0.5 < H < 1.$$

H, called the Hurst exponent, is a measure of the strength of the LTP and

$$H = 1 - \frac{\alpha}{2}$$

Defn. VI **Self-similarity:** Similar to the above definition, a process is self-similar if a property such as distribution is preserved over large scales of space and/or time

X_{mt} and $m^H X_t$ have identical distributions for all $m > 0$

Here m is a scaling factor and H is the constant Hurst exponent. Self-similarity can refer to a number of properties being preserved irrespective of scaling in space and/or time, such as variance or autocorrelation. This can provide very concise descriptions of behaviour of widely varying scales, such as the 'burstiness' of internet traffic [KMF04].

Here we show, and this is far from accepted, is that LTP is a fact of natural phenomena. LTP is seen by some to be an 'exotic' phenomenon requiring system with 'long term memory'. However, if for whatever reason systems do exhibit LTP behavior, it is important to incorporate LTP into our assumptions.

Here examine a set of proxy series listed in Table 9.1 of the previous chapter, as well as the temperature and precipitation from a sample of landscape.

10.1 Detecting LTP

One of the main operations in examining LTP are aggregates of series. Aggregates are calculated as follows. For example, given a series of numbers X, the aggregated series X_1, X_2 and X_3 is as follows.

```
> x <- seq(0, 1, by = 0.1)
> hagg(x, 1:3, sum)
```

```
[[1]]
 [1] 0.0 0.1 0.2 0.3 0.4 0.5 0.6 0.7 0.8 0.9 1.0

[[2]]
[1] 0.1 0.5 0.9 1.3 1.7

[[3]]
[1] 0.3 1.2 2.1
```

Figure 10.1 compares the two diagnostic tools. The first used previously is the plot of the ACF for successive lags. Note that the series walk, SSS and CRU decay slowly relative to IID and AR. This is consistent with Definition I of LTP in these series with high correlations at long lags.

Figure 10.2 shows a similar pattern in a different way, by plotting the correlation at lag 1 against the series aggregated at successively long time scales. The persistence of autocorrelation at higher aggregations for series walk, SSS and CRU decay over IID and AR is clear.

A plot of the logarithm of standard deviation of the simulated series against the logarithm of the level of aggregation or time scale on Figure 10.3 shows scale invariant series, as per definitions V and VI, as straight lines with a slope greater than 0.5. Random numbers form a straight line of low slope (0.5). The random walk is also a straight line of higher slope as are CRU and SSS.

Notably the slope of the AR(1) model declines with higher aggregations, converging towards the slope of the random line. This demonstrates that AR has STP as per definition V but not LTP.

Niche Modeling

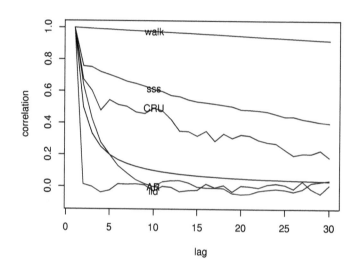

FIGURE 10.1: One way of plotting autocorrelation in series: the ACF function at lags 1 to k.

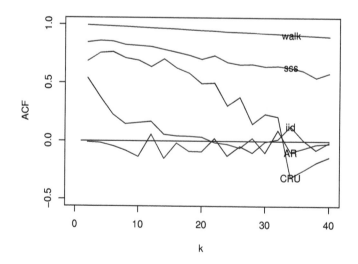

FIGURE 10.2: A second way of plotting autocorrelation in series: the ACF at lag 1 of the aggregated processes at time scales 1 to k.

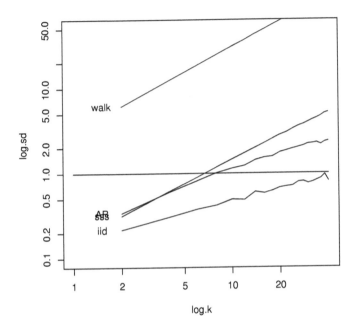

FIGURE 10.3: The log-log plot of the standard deviation of the aggregated simulated processes vs. scale k.

The implications of high self similarity or H value are most apparent in the standard error, or standard deviation of the mean. The standard error of IID series and AR series increase with the square root of aggregations. The aggregation is equivalent to sample size. Thus the usual rule for calculating standard error of the mean applies:

$$s.e. = \frac{\sigma}{\sqrt{k}}$$

Series such as the SSS, CRU and the random walk would maintain high standard errors of the mean with increasing sample size. This means that where a series has a high H, increasing numbers of data do not decrease our uncertainty in the mean very much. Alternatively, there are few effective points.

At the level of the CRU of $H = 0.95$ the uncertainty in a mean value of 30 points is almost as high as the uncertainty in a mean of a few points. It is this feature of LTP series that is of great concern where accurate estimates of confidence are needed.

Below are H estimated for all data, and estimates such as the generalized s.e. above should be used unless classical statistics can be shown to apply. On a log-log plot of standard deviation this equation is a straight line from which H can be estimated.

$$log(StDev(k)) = c + Hlog(k)$$

We can calculate the H values for the simulated series from the slope of the regression lines on the log-log plot as listed in Table 10.1. The random series with no persistence has a Hurst exponent of about 0.5. As expected the H of the AR(1) model is a low 0.67 while the SSS model we generated has an H of 0.83. The global temperatures have a high H of 0.94 and the random walk is close to one.

TABLE 10.1:
Estimates of Hurst
exponent for all
series.

	names	H
1	CRU	0.94
2	J98	0.87
3	MBH99	0.87
4	MJ03	0.92
5	CL00	0.97
6	BJ00	0.87
7	BJ01	0.84
8	Esp02	0.91
9	Mob05	0.91
10	iid	0.47
11	AR	0.66
12	walk	0.99
13	sss	0.93
14	precip	0.85
15	temp	0.93

10.1.1   Hurst Exponent

All natural series long term persistent, including temperature and precipitation all have high values of H as shown in Table 10.1. Figure 10.4 below is a log-log plot of the standard deviation of the temperature reconstructions

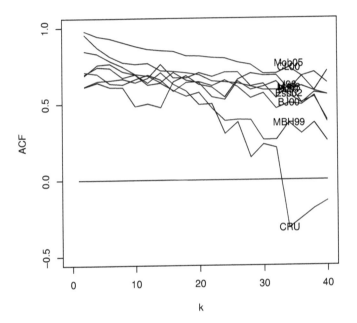

FIGURE 10.4: Lag 1 ACF of the proxy series at time scales from 1 to 40.

with respect to scale. The lines indicate highly persistent autocorrelation in all reconstructions although the slopes of the line differ slightly.

Similarly Figure 10.5 shows the lag-one ACF against scale for temperature and precipitation with the simulated series for comparison. They too show high levels of autocorrelation.

The Hurst exponents or precipitation and temperature are 0.85 to 0.93 as shown in Table 10.1. Figure 10.6 confirms that temperature and precipitation have long term persistence, shown by the straight lines with similar slope to SSS. Note the precipitation line does appear to diminish in variance at greater aggregation.

10.1.2 Partial ACF

Autocorrelation function (ACF) plots of the comparative function (IID, MA, AR, SSS) allow comparison with the autocorrelation of natural series (CRU, temperature, rainfall). Comparison of the two show the natural series are not IID but have autocorrelation properties more like combination of MA

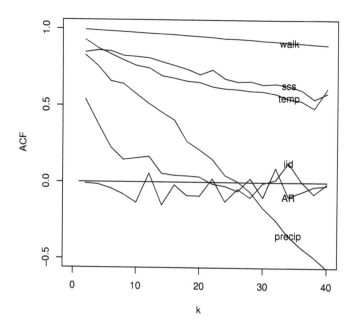

FIGURE 10.5: Lag 1 ACF of temperature and precipitation at time 1 to 40 with simulated series for comparison.

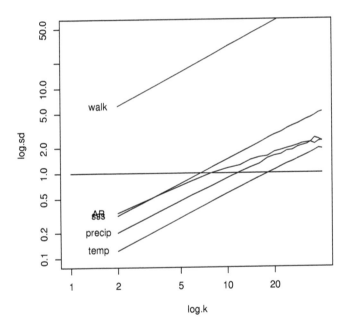

FIGURE 10.6: Log-log plot of the standard deviation of the aggregated temperature and precipitation processes at scales 1 to 40 with simulated series for comparison.

and AR or SSS series.

The ACF function in R contains another useful diagnostic option. The partial correlation coefficient is estimated by fitting autoregressive models of successively higher orders up to lag.max. Partial autocorrelations are useful in identifying the order of an autoregressive model. The partial autocorrelation of an AR(p) process is zero at lag p+1 and greater.

Figure 10.7 shows the partial correlation coefficients of the simple series, IID, MA, AR, and SSS. Figure 10.8 shows the natural series CRU, MBH99, precipitation and tempemperature. The partial correlations of the IID and the AR(1) decay rapidly. The SSS decays more slowly, as does the CRU, providing further evidence of higher order complexity of the global average temperature series. The partial correlations of the spatial temperature and precipitation series decay more quickly than CRU and appear to oscillate, indicative of moving averaging, as seen in the MA series. The partial correlation plot suggests the CRU temperatures should be modelled by an autoregressive process of at least order AR(4).

10.2   Implications of LTP

The conclusions are clear. Natural series are better modeled by SSS type models than AR(1) or IID models. The consequences are that the variance in an LTP series is greater than IID or simple AR models at all scales. Thus using IID variance estimates will lead to Type 2 errors, spurious significance, and the danger of asserting false claims.

We estimate the degree to which the confidence limits in natural series will be underestimated by assuming IID errors with LTP behavior. The normal relationship for IID data for the standard error of the mean with number of data n

$$SE[IID] = \sigma/sqrt(n)$$

has been generalized to the following form in [Kou05a]

$$SE[LTP] = \sigma/n^{1-H}$$

where the errors are IID then $H = 0.5$ and the generalized from becomes identical to the upper IID form. The increase in standard deviation for non-IID errors can be obtained by dividing and simplifying the equations above:

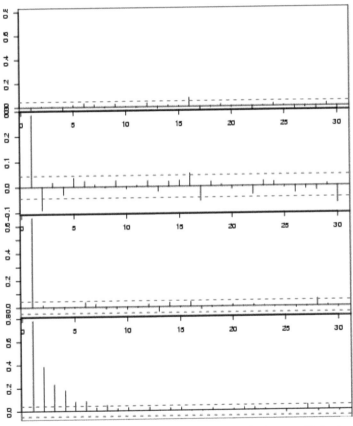

FIGURE 10.7: Plot of the partial correlation coefficient of the simple diagnostic series IID, MA, AR and SSS.

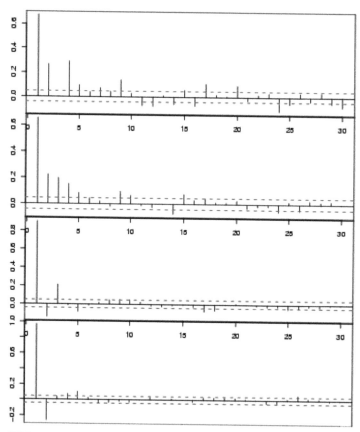

FIGURE 10.8: Plot of the partial correlation coefficient of natural series CRU, MBH99, precipitation and temperature.

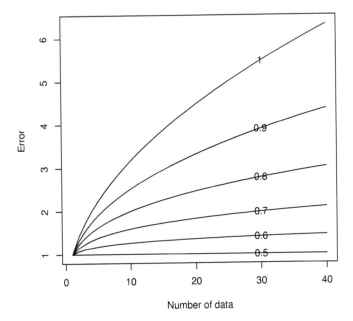

FIGURE 10.9: A: Order of magnitude of the s.d. for FGN model exceeds s.d. for IID model at different H values.

$$SE[LTP]/SE[IID] = n^{H-0.5}$$

This is plotted by n at a number of values of H in Figure 10.9. It can be seen that at the higher H values the SE[LTP] can be many times the SE[IID] (Figure 10.9).

For example, when the 30 year running mean of temperature is plotted against the CRU temperatures it can be seen that the temperature increase from 1950 to 1990 is just outside the 95% confidence intervals for the FGN model (dotted line) (Figure 10.10). The CIs for the IID model are very narrow however (dashed line).

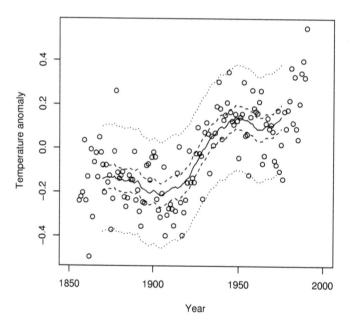

FIGURE 10.10: Confidence intervals for the 30 year mean temperature anomaly under IID assumptions (dashed line) and FGN assumptions (dotted lines).

10.3   Discussion

It is shown in [Kou06] that the use of an annual AR(1) or similar model amounts to an assumption of a preferred (annual) time scale, but there is neither a justification nor evidence for a preferred time scale in natural series. The self-similar property of natural series amounts to maximization of entropy simultaneously at all scales [Kou05b]. Thus in order for the processes that may be responsible for a series to be consistent with the second law of thermodynamics, which states that overall entropy is never decreasing, they must have LTP.

Not only do these results provide evidence that all available natural series exhibit scaling behavior, it also shows how inappropriate error models based on the classical (IID) statistical model are. In all cases the Hurst coefficient H is as high as 0.90 ± 0.07, far from 0.5. The last part of the article points out the significant implications. Clearly, niche modeling needs to be rectified in order to harmonize with this complex nature of physical processes.

However, probably the implications are even worse than described. In fact, the formula $SE[LTP]/SE[IID] = n^{(H-0.5)}$ and the relevant plot indicate the increase of uncertainty under the SSS behavior, if H is known a priori. Note that in the IID case there is no H (or $H = 0.5$ a priori) and, thus, no uncertainty about it. In the SSS case, H is typically estimated from the data, so there is more uncertainty due to statistical estimation error. This, however, is difficult (but not intractable) to quantify. And in the case of proxy data, there is additional uncertainty due to the proxy character of the data. This is even more difficult to quantify.

Chapter 11

Circularity

Another form of error more difficult to detect embodies a logical fallacy known as *petitio principii*, or the circular argument. When a conclusion is circular, usually the methodology has assumed the conclusions implicitly in the premises. In symbolic logic, the form of inference called *modus ponens* is corrupted in circular reasoning into the tautology:

If $p \Rightarrow q$ and q is true conclude q

The error here is that the truth of q is not entailed by the truth of p. One way of testing for circularity is to see if a methodology generates the same results with random data, as with the real data. This, together with the initial premise that $p \Rightarrow q$, is equivalent to setting:

$p \Rightarrow q$ and $\neg p \Rightarrow q$

If the conclusion is still true, i.e. the same results are achieved using random numbers instead of the supposed 'signal', then it is highly likely the model entails the result q irrespective of the conditions.

11.1   Climate prediction

Circularity is illustrated on the climate reconstruction model used in a previous chapter, where randomly generated sequences are used to develop predictions of past climate. We examine the potential for circularity via the selection of series, or 'cherry picking' for series highly correlated with temperature.

The process through which temperature reconstructions are developed from tree-rings is to collect a number of tree-ring widths or densities from cores of large trees older than the instrumental temperature record. These series are

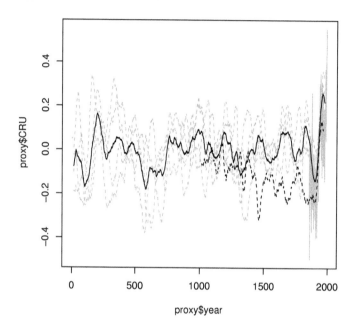

FIGURE 11.1: A reconstruction of temperatures generated by summing random series that correlate with temperature.

normalized to reduce the effect of promiscuous early growth.

The series are then tested for correlation with temperature. Those that are best correlated with the instrumental temperature record are selected, either prior to analysis or through differential weighting in statistical techniques. The selected series are then calibrated and combined, either by averaging or another technique.

Figure 11.1 shows reconstructed temperature anomalies over 2000 years, Dashed line being a published reconstruction of temperatures [MBH98], the individual random series in grey and the climate reconstruction from random series in black. Note the similarity in the overall 'hockey stick' shape.

11.1.1   Experiments

The r^2 is ubiquitous for relating correlation of variables, and values above 0.1 would usually barely be a significant correlation. Other measures such as RE statistic are claimed to be superior to the r^2 statistic [WA06] due to sensitivity to mean values, i.e. if the mean of the predicted series is close to the test points.

A number of parameters are varied while repeating the generation of random series and temperature reconstruction. We then determine the effect of aspects of the generalized methodology on model skill by dropping each in turn:

- different types of series (IID, alternating means and fractional differencing),

- dropping requirements for positive slope,

- dropping requirements for positive correlation, and

- dropping requirements for calibration with inverse linear model.

Once again we generate 100 sequences to develop a reconstruction, and repeat this 10 times to estimate the mean and standard deviation of statistics for verifying skill on the following:

- training and test data, for

- raw and Gaussian filtered CRU temperatures.

Table 11.1 shows the results of these experiments. All except two of the variations produced results indicating significant skill for the random model. The two that didn't indicate skill were the random sequences using IID errors, and the methodology that drops the $r^2 > 0.1$ requirement. With IID series, none of the 100 random sequences were found to be significant, so no reconstructions could be developed. Without the correlation for r^2, the quality of the reconstructions dropped considerably, as shown by the lower r^2 values in the skill of the final reconstructions.

Dropping requirements for positive slope and for calibration of selected series did not affect skill of reconstructions greatly. In other words, trees with negative correlations with temperatures could equally be used, as they could be regarded as negatively sensitive to temperature, and the calibration stage coerces them to a positive orientation.

These results show the elements of the procedure that can lead to spurious determination of skill. Firstly, random series with high autocorrelation are necessary to generate a high proportion of spurious correlations (about 20%). As we know from previous chapters this is the case with natural series. Therefore it is likely that series from individual tree-ring records would display the same high levels of autocorrelation.

TABLE 11.1:   Correlations of random model resulting from
out of range validation of different experiments.

	inCRUr	exCRUr	infCRUr	exfCRUr	n
All - IID	0.00	0.00	0.00	0.00	0.00
All - altmeans	0.50	0.49	0.87	0.87	16.00
All - fracdiff	0.51	0.49	0.88	0.90	20.00
Arb. slope	0.50	0.53	0.90	0.90	42.00
Arb. r^2	0.15	0.15	0.32	0.35	51.00
Uncalibrated	0.47	0.07	0.90	0.90	20.00

Secondly, the selection of trees for correlation with temperatures is also a factor necessary for spurious attributions of skill. Thus, selection of trees for temperature sensitivity can be a major factor in generating apparently skillful reconstructions of climate from random sequences. This demonstrates that 'cherry picking' is a major cause of spurious attribution of statistical skill to models.

The reconstruction skill is relatively insensitive to the other factors: the calibration and requirement for positive slope.

One can conclude that the overall pattern of the model – the hockey-stick shape – can be produced with randomly generated data and two factors: high levels of autocorrelation resulting in spurious regression, and 'cherry picking' only those random series correlating with temperatures. It is due to these factors that the reconstructions of temperature show increasing 20th century values, and the average of the series reverts to the mean value of the random numbers outside the range of the calibration temperatures.

This figure has great similarity to other published reconstructions, particularly the prominent hockey-stick shape, the cooler temperatures around the 1500s and the Medieval Warm Period around the 1000s [BO02]. The degree to which these constructions are based on series selected specifically for their correlation with temperature should raise questions. Circular reasoning undermines the conclusions that the hockey-stick shape is real, and opens studies to the criticism that the result is an artifact of 'cherry picking' series for hockey stick shapes.

In this situation, the types of errors, the selection for sequences that calibrate with r^2 effectively, constitute assumptions relevant to existing climate reconstructions. Alternative approaches that would render the method immune from claims of circular reasoning are:

prove first that errors are IID. If IID the series would be very unlikely to have members that correlate significantly with the CRU temperatures; else do not select the individual series by calibration, 'cherry pick' or otherwise calibrate individual series with temperature. Rather, apply model developed from first principles to all available series.

If neither of the above is avoidable, then estimate the expected null result from a Monte Carlo simulation as illustrated in Figure 11.1. Then test results and claim only significant deviations from the Monte Carlo simulation are significant.

Comparison of the actual reconstruction in Figure 11.1 indicates the only significant deviation from the random reconstruction is during the period preceding the beginning of the last century, and may indicate a real pattern. The existence of this period, known as the Little Ice Age (LIA) as well as 20th century warming are well supported by other evidence. However, these results demonstrate that conclusions of recent warming using this methodology would be circular.

11.2 Lessons for niche modeling

The result is relevant to the selection of variables for modeling.

- We have shown that the most important variables are highly autocorrelated (Chapter 10). Therefore option 1 is not an option. As natural correlation is a form of long term persistence this is unavoidable by aggregation or other means.

- Developing species niche models with a small set of relevant variables from first principles has been achieved using physiological characteristics of species. However these models require a great deal of research into species characteristics.

- For modeling poorly understood species it is necessary to select some variables based on their correlation (Chapter 4) and this means 'cherry picking' is unavoidable.

- Other forms of error such as non-linearity and bias are also prevalent and unavoidable.

Thus it would follow that a Monte Carlo approach is inevitable for developing reliable species niche models in 2 dimensions. The procedure would be as follows:

- Generate a set of random variables with the same noise and autocorrelation properties as variables such as temperature and rainfall.

- Develop a null model of species distributions first on those variables, and predict distribution.

- Generate the niche model on real variables, using the benchmark above for rejection.

- Predict the distribution.

- Compare the spatial distribution of the real results with the distribution from the random variables.

- Claim results only if significance is higher than the random variables, and the distribution differs in obvious ways from the random prediction.

These results suggest such an approach is necessary in order to avoid the forms of error examined here.

Chapter 12

Fraud

Can the fabrication of research results be prevented? Can peer review be augmented with automated checking? These questions become more important with the increase in automated submission of data to portals. The potential usefulness of niche modeling to detecting at least some forms of either intentional or unintentional 'result management' is examined here.

Benford's Law is a postulated distributional relationship on the frequency of digits [Ben38]. It states that the distribution of the combination of digits in a set of random data drawn from a set of random distributions follows the log relationship for each of the digits, as shown in the Figure 12.1. Benford's Law, actually more of a conjecture, suggests the probability of occurrence of a sequence of digits d is given by the equation:

$Prob(d) = log_{10}(1 + \frac{1}{d})$

For example, the probability of the sequence of digits 1,2,3 is given by $log_{10}(1 + \frac{1}{123})$. The frequency of digits can deviate from the law for a range of reasons, mostly to do with constraints on possible values.

Deviations due to human fabrication or alteration of data have been shown to be useful for detecting fraud in financial data [Nig00]. Although Benford's law holds on the first digit of some scientific datasets, particularly those covering large orders of magnitude, it is clearly not valid for data such as simple time series where the variance is small relative to the mean. As a simple example, the data with a mean of 5 and standard deviation of 1 would tend to have leading digits around 4 or 5, rather than one.

Despite this, it is possible that subsequent digits may conform better. A recent experimental study suggested the second digit was a much more reliable indicator of fabricated experimental data [Die04]. Such a relationship would be very useful on time series data as generated by geophysical data.

This paper reports some tests of the second digit frequency as a practical methodology for detecting 'result management' in geophysical data. It also illustrates a useful generalization of niche modeling as a form of prediction based on models of statistical distribution.

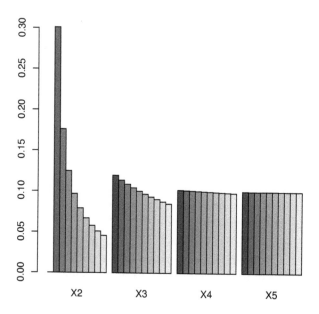

FIGURE 12.1: Expected frequency of digits 1 to 4 predicted by Benford's Law.

12.1 Methods

Here we looked at histograms and statistics, Chi square tests of deviation from Benford's Law and uniform distribution, and normed distance of digit frequency for the first and second digit. We also generate a plot of the chi square values of digit frequencies in a moving window over a time series plot. This is useful for diagnosing what parts of a series deviate from the expected digit distribution.

Four datasets are tested. These are:

- simulated dataset composed of random numbers and fabricated data,

- Climate Research Unit - CRU - composed of global average monthly temperatures from meteorological stations from 1856 to present,

- tree ring widths drawn from the WDCP paleoclimatology portal, and

- tidal height dataset, collected both by hand recording and instrumental reading.

12.1.1 Random numbers

A set of numbers with an IID distribution were generated. I then fabricated data to resemble the random numbers. Below are the result for the first and second digit distribution with fits on a log-log plot and residuals. The plots show that while the first digit deviates significantly, the second does not (Figure 12.2).

A number of statistics from digit frequency were calculated. The first two are from Nigrini [Nig00]. These include:

df indicating management of results up or down as might occur with rounding of results financial records up or down,

z score,

chi-square value, and

distance of distribution from expected.

On examining the probability of conformance of digit frequency with Benford's Law (P) on the random data, the first digit appears mildly deviant, while the second digit is not (Figure 12.3). The significant deviation of the second digit is correctly identified on the fabricated data as well (Table 12.1).

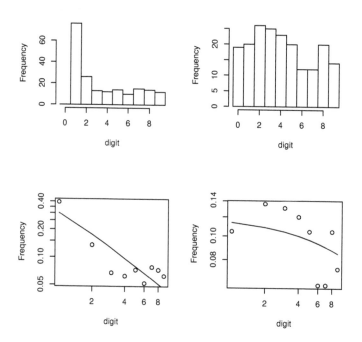

FIGURE 12.2:   Digit frequency of random data.

Another way of quantifying deviation is to sum the norm of the difference between the expected and observed frequencies for each digit (D). The value for D on the second digit of the fabricated data is much higher than the value for random data. The difference between random and fabricated data in first digit is less clear. These results give one confidence that statistical tests of the second digit frequency can detect fabrication in datasets. The second digit method appears more useful on these types of geophysical data than Nigrini's method (df, z), developed primarily for detecting results management in financial data.

Figure 12.4 is the result on time series data. The solid line is the significance of the second digit where the p is calculated on the moving window of size 50. The dashed line is a benchmark level of probability below which indicates deviation from Benford's Law distribution.

Figure 12.5 shows the differenced series. The line dipping below the dashed line shows differenced fabricated data is detected by the test of the distribution of the second digit against Benford's Law, although the results are less clear.

The simulated data consist of a random sequence on both sides, and fabricated data in the center. In the figure above, the fabricated region is clearly

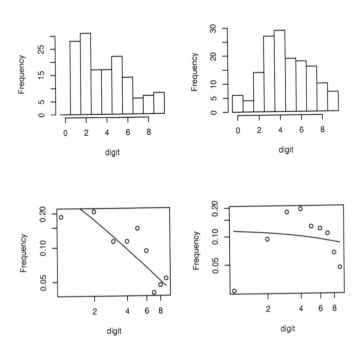

FIGURE 12.3: Digit frequency of fabricated data.

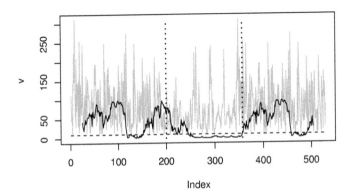

FIGURE 12.4: Random data with section of fabricated data inserted in the middle.

TABLE 12.1: Statistics from digit frequency of
random and fabricated data: df - management up or
down, z score, chi-square value, and distance of
distribution from expected.

	description	df	z	P	D
1	random 1st digit	1.17	25.31	0.01	0.31
2	random 2nd digit	1.17	25.31	0.46	0.20
3	fabricated 1st digit	0.60	11.48	0.01	0.30
4	fabricated 2nd digit	0.60	11.48	0.00	0.48

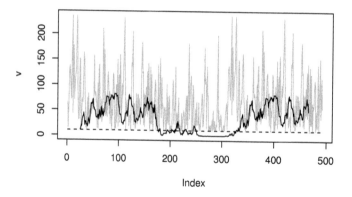

FIGURE 12.5: The same data above differenced with lag one.

detected in both the original series and the differenced series when the red
line falls below the dashed line on both raw and differenced plots.

12.1.2 CRU

Global average temperature files at CRU are updated on a monthly basis
to include the latest month with various automated procedures including a
method of variance adjustment [JNP+99]. The annual data were examined
for digit frequency below (Figures 12.6). They do not appear to show management of results (Table 12.2).

```
1856 -0.384 -0.457 -0.673 -0.344 -0.311 ...
   1856      16     17     16     16     15     ...
```

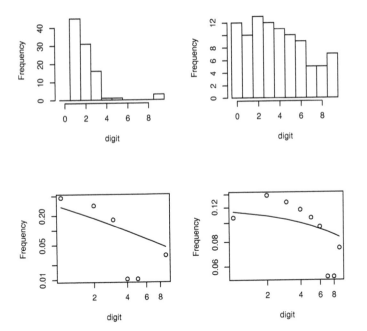

FIGURE 12.6: First and second digit frequency of CRU data.

TABLE 12.2: Indices of first and second digit frequency in CRU global temperatures.

	description	df	z	P	D
1	1st digit	−0.46	−7.09	0.00	0.52
2	2nd digit	−0.46	−7.09	0.87	0.18

12.1.3 Tree rings

The tree ring data is freely available from the World Center for Paleoclimatology portal at http://www.ncdc.noaa.gov/paleo/data.html. A file called pa.rwl was chosen virtually at random for analysis. The header information for this file gives the location of the tree ring samples, the species, the collector, and time and location information (Figure 12.7 and 12.8). The files show signs of results management, possibly due to the hand transcription of measurement data (Table 12.3)

```
120-1   1 ELY LAKE
120-1   2 PA, USA        EASTERN HEMLOCK     1391   4146N07550
1891 1973 120-1   3 SWAIN      06 17 74
120106  1911     149     145     139     282     322     224     219     230     153
```

TABLE 12.3: First and second digit indices for tree ring data.

	description	df	z	P	D
1		−0.58	−28.85	0.00	0.52
2		−0.58	−28.85	0.00	0.18

12.1.4 Tidal Gauge

The tidal height sets were obtained from John Hunter (john.hunter@utas.edu.au, Hunter et.al. 2003) The manually recorded dataset porta40.txt is a digitized versions of sea level data hand collected in imperial units (feet and inches) by Thomas Lempriere at Port Arthur, Tasmania, Australia. They cover the period 1840 to 1842, but are incomplete for 1840 [HCD03].

The second file called present_dat_gmt is the full (1184-day) data-set from the Port Arthur tide gauge from 24/6/1999 to 20/9/2002 collected with an Aquatrak 4100 with Vitel WLS2 logger. The 2nd digit frequencies are shown

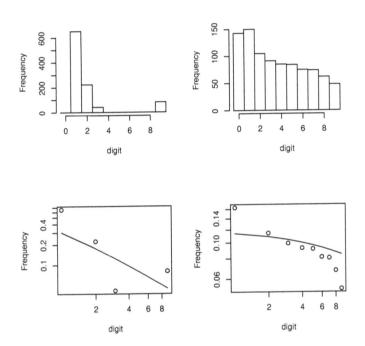

FIGURE 12.7: Digit frequency of tree-ring data.

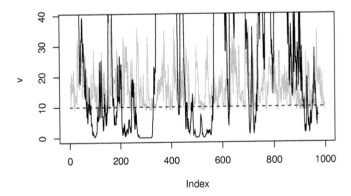

FIGURE 12.8: Digit significance of tree-ring series.

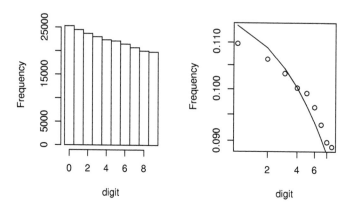

FIGURE 12.9: Digit frequency of tidal height data, instrument series.

on Figures 12.9 and 12.10.

Figure 12.9 shows the distribution of the second digit in the hand collected data deviates significantly from Benford's Law predictions due to elevated frequencies of 0's and 5's. This may have been due to human 'rounding error' biasing data towards either whole or half feet. The proof of some form of results management in the hand coded dataset is shown in the table below. The Chi2 test on the second digit shows significant deviance from the expected Benford's distribution in both datasets. However the distance measure appears to discriminate between the two sets, with the instrumental set giving a D2 value of 0.03 and the hand-coded set a value of 0.33 (Table 12.4).

12.1.5 Tidal gauge - hand recorded

TABLE 12.4: Digit frequency of tidal data by hand and by instrument.

	description	df	z	P	D
1	instrument, 2nd digit	−0.10	−77.05	0.00	0.03
2	hand recorded, 2nd digit	0.17	9.88	0.00	0.25

The figure 12.11 shows the Chi2 value dropping below the green line in a number of regions throughout the series. John Hunter communicated that

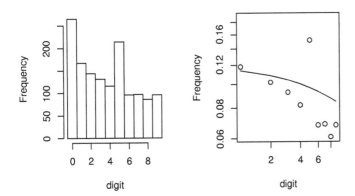

FIGURE 12.10:   Digit frequency of tidal height data - hand recorded.

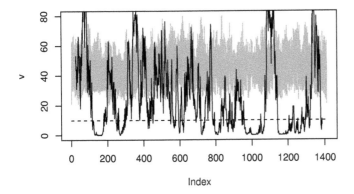

FIGURE 12.11:   Digit significance of hand recorded set along series.

he believed different people were responsible for collection of data throughout the period, which may explain the transitions in the graph.

12.2   Summary

   The 2nd digit distribution discriminated randomly generated and fabricated data. A moving window analysis correctly identified the location of the fabricated data in a time series. The digit distribution of the tidal gauge dataset collected by instruments showed significant deviations from Benford's Law. In particular, the tidal gauge dataset collected by hand showed evidence of rounding of results to whole or half feet. The instrumental dataset is very large (n=28,179) and this could account for the sensitivity of the Chi2 to deviation. Another explanation for the deviation may be the conjecture that Benford's Law is not exactly accurate. These results show that the distributions of the 2nd digit can be used to detect and diagnose 'results management' on geophysical time series data. These methods could assist in quality control of a wide range of datasets, particularly in conjunction with automated data archival processes.

References

[AM96] M. Austin and J. Meyers. Current approaches to modelling the environmental niche of eucalypts: Implication for management of forest biodiversity. *Forest Ecology And Management*, 85(1-3):95–106, 1996.

[AMN⁺03] R. B. Alley, J. Marotzke, W. D. Nordhaus, J. T. Overpeck, D.M. Peteet, R. A. Pielke Jr., R. T. Pierrehumbert, P. B. Rhines, T. F. Stocker, L. D. Talley, and J. M. Wallace. Abrupt climate change. *Science*, 299(5615), 2003.

[AMS⁺97] R. Alley, P. Mayewski, T. Sowers, M. Stuiver, K. Taylor, and P. Clark. Holocene climatic instability: A prominent, widespread event 8200 yr ago. *Geology*, 25(6), 1997.

[Aus87] M. P. Austin. Models for the analysis of species' response to environmental gradients. *Plant Ecology (Historical Archive)*, 69(1 - 3):35–45, 1987.

[Ben38] F. Benford. The law of anomalous numbers. *Proceedings of the American Philosophical Society*, 78(4):551–572, 1938.

[BHP05] Linda J. Beaumont, Lesley Hughes, and Michael Poulsen. Predicting species distributions: use of climatic parameters in bioclim and its impact on predictions of species' current and future distributions. *Ecological Modelling*, 186:251–270, 2005.

[BO02] K.R. Briffa and T.J. Osborn. Paleoclimate. Blowing hot and cold. *Science*, 295(5563):2227–2228, Mar 2002.

[Bri00] Keith R. Briffa. Annual climate variability in the holocene: interpreting the message of ancient trees. *Quaternary Science Reviews*, 19:87–105, 2000.

[CAB⁺02] M. E. Conkright, J. I. Antonov, O. Baranova, T.P. Boyer, H.E. Garcia, R. Gelfeld, D.D. Johnson, R.A. Locarnini, P.P. Murphy, T.D. O'Brien, I. Smolyar, and C. Stephens. World ocean database 2001. NOAA Atlas 42, NESDIS, U.S. Government Printing Office, Washington, D.C., 167 pp., 2002.

[CGW93] G. Carpenter, A. N. Gillison, and J. Winter. Domain: a flexible modelling procedure for mapping potential distributions of plants and animals. *Biodiversity and Conservation*, 2:667–680, 1993.

[CTLT00] T.J. Crowley, J. Thomas, T. Lowery, and S. Thomas. How warm was the medieval warm period? *AMBIO: A Journal of the Human Environment 2000*, 29:51–54, 2000.

[Die04] A. Diekmann. Not the first digit! using benford's law to detect fraudulent scientific data. Technical report, Swiss Federal Institute of Technology, Zurich, 2004.

[DKD+04] R. D'Arrigo, R. Kaufmann, N. Davi, G. Jacoby, C. Laskowski, R. Myneni, and P. Cherubini. Thresholds for warming-induced growth decline at elevational tree line in the yukon territory, canada. *Global Biogeochemical Cycles*, 18(3), 2004.

[ECS02] Jan Esper, Edward R. Cook, and Fritz H. Schweingruber. Low-frequency signals in long tree-ring chronologies for reconstructing past temperature variability. *Science*, 295(5563):2250–2253, 2002.

[EGA+06] J. Elith, C. H. Graham, R. P. Anderson, M. Dudik, S. Ferrier, A. Guisan, R. J. Hijmans, F. Huettmann, J. R. Leathwick, A. Lehmann, J. Li, L. G. Lohmann, B. A. Loiselle, G. Manion, C. Moritz, M. Nakamura, Y. Nakazawa, J. M. Overton, T. A. Peterson, S. J. Phillips, K. Richardson, R. Scachetti-Pereira, R. E. Schapire, J. Soberon, S. Williams, M. S. Wisz, and N. E. Zimmermann. Novel methods improve prediction of species distributions from occurrence data. *Ecography*, 29(2):129–151, April 2006.

[FHT98] J. Friedman, T. Hastie, and R. Tibshirani. Additive logistic regression: a statistical view of boosting, 1998.

[FK03] Oren Farber and Ronen Kadmon. Assessment of alternative approaches for bioclimatic modeling with special emphasis on the mahalanobis distance. *Ecological Modelling*, 160:115–130, 2003.

[HCD03] J. Hunter, R. Coleman, and Pugh D. The sea level at port arthur, tasmania, from 1841 to the present. *Geophysical Research Letters*, 30(7):54–1 to 54–4, 2003.

[HCP+05] R.J. Hijmans, S.E. Cameron, J.L. Parra, P.G. Jones, and A. Jarvis. Very high resolution interpolated climate surfaces for global land areas. *International Journal of Climatology*, 25:1965–1978, 2005.

[HDT+03] M. Hansen, R. DeFries, J.R. Townshend, M. Carroll, C. Dimiceli, and R. Sohlberg. 500m modis vegetation continuous fields, 2003.

[Hec82] Paul Heckbert. Color image quantization for frame buffer display. In *SIGGRAPH '82: Proceedings of the 9th annual conference on Computer graphics and interactive techniques*, pages 297–307, New York, NY, USA, 1982. ACM Press.

[HM82] J.A. Hanley and B.J. McNeil. The meaning and use of the area under a receiver operating characteristic (roc) curve. *Radiology*, 143:29–36, 1982.

[Hut58] G.E. Hutchinson. Concluding remarks, cold spring harbour symposia on quantitative biology. In *Cold Spring Harbor Symposia on Quantitative Biology.*, volume 22, pages 415–427, 1958.

[JMF99] A. K. Jain, M. N. Murty, and P. J. Flynn. Data clustering: a review. *ACM Comput. Surv.*, 31(3):264–323, 1999.

[JNP+99] P.D. Jones, M. New, D.E. Parker, S. Martin, and I.G. Rigor. Surface air temperature and its variations over the last 150 years. *Reviews of Geophysics*, 37:137–199, 1999.

[JS00] L. Joseph and D. Stockwell. Temperature-based models of the migration of swainson's flycatcher (myiarchus swainsoni) across south america: A new use for museum specimens of migratory birds. *Proceedings of the National Academy of Sciences Philadelphia*, 150, 2000.

[JW99] S. T. Jackson and C. Weng. Late quaternary extinction of a tree species in eastern north america. *Proc Natl Acad Sci U S A*, 96(24), 1999.

[Ker05] Richard A. Kerr. GLOBAL WARMING: Millennium's Hottest Decade Retains Its Title, for Now. *Science*, 307(5711):828a–829, 2005.

[KHO+00] J.J. Kineman, D.A. Hastings, M.A. Ohrenschall, J. Colby, D.C. Schoolcraft, J. Klaus, J. Knight, L. Krager, P. Hayes, K. Oloughlin, P. Dunbar, J. Ikleman, C. Anderson, J. Burland, J. Dietz, H. Fisher, A. Hannaughan, M. Kelly, S. Boyle, M. Callaghan, S. Delamana, L. Di, K. Gomolski, D. Green, S. Hochberg, W. Holquist, G. Johnson, L. Lewis, A. Locher, A. Mealey, L. Middleton, D. Mellon, L. Nigro, J. Panskowitz, S. Racey, B. Roake, J. Ross, L. Row, J. Schacter, and Weschler. P. Global ecosystems database version ii: Database, user's guide, and dataset documentation. Two CDROMs and publication on the World Wide Web, 2000.

[KMF04] T. Karagiannis, M. Molle, and M. Faloutsos. Long-range dependence: Ten years of internet traffic modeling. *Internet Computing*, 8(5):57–64, 2004.

[Kou02] D. Koutsoyiannis. The hurst phenomenon and fractional gaussian noise made easy. *Hydrological Sciences Journal*, 47(4):573–596, 2002.

[Kou05a] D. Koutsoyiannis. Nonstationary versus scaling in hydrology. *Journal of Hydrology*, in press, 2005.

[Kou05b] Demetris Koutsoyiannis. Uncertainty, entropy, scaling and hydrological stochastics, 2, time dependence of hydrological processes and time scaling. *Hydrological Sciences Journal*, 50(3):405–426, 2005.

[Kou06] Demetris Koutsoyiannis. An entropic-stochastic representation of rainfall intermittency: The origin of clustering and persistence. *Water Resources Research*, 42, 2006.

[Len00] J.J. Lennon. Red-shifts and red herrings in geographical ecology. *Ecography*, 23:101–113, 2000.

[LL58] L. D. Landau and E. M. Lifshitz. *Statistical Physics*. Pergamon, New York, 1958.

[Ll96] C. Loehle and D. leBlanc. Model-based assessments of climate change effects on forests: a critical review. *Ecological Modelling*, 90, 1996.

[LLK⁺00] M. Lindholm, H. Lehtonen, T. Kolstrom, J. Merilainen, M. Eronen, and M. Timonen. Climatic signals extracted from ring-width chronologies of scots pines from the northern, middle and southern parts of the boreal forest belt in finland. *Silva Fennica*, 34:317–330, 2000.

[Loe00] C. Loehle. Forest ecotone response to climate change: sensitivity to temperature response functional forms. *Canadian Journal of Forest Research*, 30:1632–1645, 2000.

[Loe03] C. Loehle. Competitive displacement of trees in response to environmental change or introduction of exotics. *Environ Manage*, 32(1), 2003.

[MBH98] M. E. Mann, R. S. Bradley, and M. K. Hughes. Global scale temperature patterns and climate forcing over the last six centuries. *Nature (London)*, 392(23), 1998.

[MGLM⁺04] B. Martrat, J. O. Grimalt, C. Lopez-Martinez, I. Cacho, F. J. Sierro, J. A. Flores, R. Zahn, M. Canals, J. H. Curtis, and D. A.

Hodell. Abrupt temperature changes in the western mediterranean over the past 250,000 years. *Science*, 306(5702), 2004.

[MHCE02] M.E. Mann, M.K. Hughes, E.R. Cook, and J. Esper. Tree-ring chronologies and climate variability. *Science*, 296(5569):848–849, 2002.

[Mun75] J.R. Munkres. *Topology: a first course*. Prentice-Hall, Inc., 1975.

[Nig00] M. Nigrini. Digital analysis using benford's law: Tests statistics for auditors. Technical report, Global Audit Publications, Vancouver, Canada, 2000.

[Nix86] H. Nix. A biogeographic analysis of australian elapid snakes. In R. Longmore, editor, *Atlas of Australian Elapid Snakes*, volume 8, pages 4–15. Australian National University, 1986.

[PAS06] Steven J. Phillips, Robert P. Anderson, and Robert E. Schapire. Maximum entropy modeling of species geographic distributions. *Ecological Modelling*, 190:231–259, 2006.

[Pet03] A. T. Peterson. Predicting the geography of species' invasions via ecological niche modeling. *Quarterly Review Of Biology*, 78(4), 2003.

[PJR+99] J. Petit, J. Jouzel, D. Raynaud, N. Barkov, J. Barnola, I. Basile, M. Bender, J. Chappellaz, M. Davis, G. Delaygue, M. Delmotte, V. Kotlyakov, M. Legrand, V. Lipenkov, C. Lorius, L. Pepin, C. Ritz, E. Saltzman, and M. Stievenard. Climate and atmospheric history of the past 420,000 years from the vostok ice core, antarctica. *Nature*, 399(6735), 1999.

[PW93] S. Portnoy and M.F. Willson. Seed dispersal curves: Behavior of the tail of the distribution. *Evolutionary Ecology*, 7(1):25–44, 1993.

[PY03] C. Parmesan and G. Yohe. A globally coherent fingerprint of climate change impacts across natural systems. *Nature*, 421(6918), 2003.

[R D05] R Development Core Team. *R: A language and environment for statistical computing*. R Foundation for Statistical Computing, Vienna, Austria, 2005.

[RFMCI99] G.H. Rodda, T.H. Fritts, M.J. McCoid, and E.W. Campbell III. An overview of the biology of the brown treesnake (boiga irregularis), a costly introduced pest on pacific islands. In Y. Rodda, G.H. Sawai, D. Chiszar, and Tanaka H., editors, *Problem snake management : the habu and the brown treesnake*, pages 44–80. Cornell University Press, Ithaca, NY. 534p., 1999.

[RPH⁺03] T. L. Root, J. T. Price, K. R. Hall, S. H. Schneider, C. Rosen-zweig, and J. A. Pounds. Fingerprints of global warming on wild animals and plants. *Nature*, 421(6918), 2003.

[SB03] Willie Soon and Sallie Baliunas. Proxy climatic and environmental changes of the past 1000 years. *Climate Research*, 23:89–110, 2003.

[SCF⁺01] M. Scheffer, S. Carpenter, J. A. Foley, C. Folke, and B. Walker. Catastrophic shifts in ecosystems. *Nature*, 413(6856), 2001.

[SP99] D. Stockwell and D. Peters. The garp modelling system: problems and solutions to automated spatial prediction. *International Journal Of Geographical Information Science*, 13(2):143–158, 1999.

[SP02a] D. R. B. Stockwell and A. T. Peterson. *Controlling bias during predictive modeling with museum data. In (, editors). Island Press.*, pages 537–546. Island Press, Washington D.C., 2002.

[SP02b] D. R. B. Stockwell and A. T. Peterson. Effects of sample size on accuracy of species distribution models. *Ecological Modelling*, 148(1), 2002.

[Sto06] David R. B. Stockwell. Improving ecological niche models by data mining large environmental datasets for surrogate models. *Ecological Modelling*, 192:188–196, 2006.

[TBAL04] W. Thuiller, L. Brotons, M. B. Araujo, and S. Lavorel. Effects of restricting environmental range of data to project current and future species distributions. *Ecography*, 27(2), 2004.

[TTR⁺04] J. Thomas, M. Telfer, D. Roy, C. Preston, J. Greenwood, J. Asher, R. Fox, R. Clarke, and J. Lawton. Comparative losses of british butterflies, birds, and plants and the global extinction crisis. *Science*, 303(5665), 2004.

[TWC⁺04] C. Thomas, S. Williams, A. Cameron, R. Green, M. Bakkenes, L. Beaumont, Y. Collingham, B. Erasmus, M. de Siqueira, A. Grainger, L. Hannah, L. Hughes, B. Huntley, A. van Jaarsveld, G. Midgley, L. Miles, M. Ortega-Huerta, A. Peterson, and O. Phillips. Biodiversity conservation - uncertainty in predictions of extinction risk - effects of changes in climate and land use - climate change and extinction risk - reply. *Nature*, 430(6995), 2004.

[Uni] Climate Research Unit. Global average temp. 1856 to 2005.

[vSZJ⁺04] H. von Storch, E. Zorita, J. M. Jones, Y. Dimitriev, F. Gonzalez-Rouco, and S. F. Tett. Reconstructing past climate from noisy data. *Science*, 306(5696), 2004.

[WA06] Eugene R. Wahl and Caspar M. Ammann. Robustness of the mann, bradley, hughes reconstruction of surface temperatures: Examination of criticisms based on the nature and processing of proxy climate evidence. *Climate Change*, in press, 2006.

[Whi04] A. Whitney. K programming language, 2004.

Index